T0039155

Interactions of Yeasts, Moulds, and Antifungal Agents

Gerri S. Hall
Editor

Interactions of Yeasts, Moulds, and Antifungal Agents

How to Detect Resistance

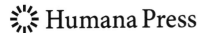 Humana Press

Editor
Gerri S. Hall, Ph.D., D(ABMM), F(AAM)
Section of Clinical Microbiology
Department of Clinical Pathology
Cleveland Clinic
Cleveland, OH 44195, USA
hallg@ccf.org

ISBN 978-1-58829-847-8   e-ISBN 978-1-59745-134-5
DOI 10.1007/978-1-59745-134-5
Springer New York Dordrecht Heidelberg London

Library of Congress Control Number: 2011941443

Printed on acid-free paper

Humana Press is part of Springer Science+Business Media (www.springer.com)

I would like to dedicate this to my husband, James O. Hall, DPM and my son James Joseph (JJ) Hall, and my secretary and dear friend Faith Cumberledge. Thanks for your support and encouragement.

Preface

The incidence of fungal infections continues to increase in hospitalized patients. *Candida* spp. have become a significant cause of bloodstream infections (BSIs) in immunocompromised and immunocompetent patients. *Candida albicans* no longer is the cause of all of these fungemia cases, but rather only about 50% of BSIs are caused by *C. albicans*; the remainder are caused by other species of *Candida* to include *C. glabrata, C. tropicalis, C. parapsilosis,* and others. Not all of these yeast responsible for fungemia have a 100% predictable response to antifungal agents. *Candida* spp. in addition can be involved in a wide spectrum of infections from candidal vulvovaginitis to postsurgical wound infections, endophthalmitis, keratitis, endocarditis, and a host of other infections. Molds like *Aspergillus* spp., *Fusarium* spp., and *Pseudallescheria boydii* are responsible for pulmonary infections in immunocompromised hosts, including transplant patients, diabetics, and patients on long-term steroids. Dermatophyte infections remain one of the most communicable infectious diseases in the world.

The number of antifungal agents has increased so that there are choices and one drug does not have to be used for all fungal infections. There is a variable response of each yeast or mold to the antifungal agents. Some are always susceptible; others are intrinsically resistant. As more of these newer agents are used, resistance has begun to emerge just as it has for bacteria. To accommodate these changes, in vitro fungal susceptibility testing is being requested more and more. The manufacturing of manual and automated methods for performing an in vitro susceptibility test has increased, and more laboratories are performing yeast susceptibilities in house, rather than sending these out to reference laboratories. Even performance of in vitro mold susceptibilities are not as uncommonly done as was once the case.

This text has been designed to cover the topic of antifungal agents and resistance detection in fungal organisms, both yeasts and molds. One chapter is devoted to a description of the most used antifungal agents, including those that are given systemically, orally, and topically. Three chapters give information on the methods that

vii

can be used for performing in vitro susceptibility tests for yeasts and molds, and the dermatophytes. The clinical utility of these in vitro tests is well described in one chapter of this text. A chapter on the usual patterns of susceptibility for common yeasts and molds is included as a reference tool for the laboratorian and the clinician. The authors hope that you will find this text useful in determining when and how in vitro testing might be done and instances where it need not be performed due to intrinsic resistances among the fungi.

Cleveland, OH, USA   Gerri S. Hall

Acknowledgments

I would like to acknowledge all of the authors who have contributed to this textbook. They are the experts in the area of laboratory and clinical diagnosis of fungi and have taught me so much throughout the years I have known them.

Contents

Contributors

David V. Chand, MSE, M.D. Division of Pediatric Infectious Diseases and Rheumatology, Department of Pediatrics, Rainbow Babies & Children's Hospital/University Hospitals of Cleveland, Cleveland, OH, USA

Daniel J. Diekema, M.D. Division of Infectious Diseases, Department of Internal Medicine, University of Iowa Carver College of Medicine, Iowa City, IA, USA

University of Iowa College of Public Health, Iowa City, IA, USA

Annette W. Fothergill, M.A., M.B.A., MT(ASCP), CLS(NCA) Fungus Testing Laboratory, Department of Pathology, University of Texas Health Science Center, San Antonio, TX, USA

Mahmoud A. Ghannoum, M.Sc., Ph.D. Center for Medical Mycology, University Hospitals of Cleveland/Case Western Reserve University, Cleveland, OH, USA

James O. Hall, D.P.M. Section of Podiatric Medicine, Cleveland Clinic, Cleveland, OH, USA

Gerri S. Hall, Ph.D. Section of Clinical Microbiology, Department of Clinical Pathology, Cleveland Clinic, Cleveland, OH, USA

Elizabeth Neuner, Pharm.D. Pharmacy Department, Cleveland Clinic, Cleveland, OH, USA

Michael A. Pfaller, M.D. Division of Clinical Microbiology, Department of Pathology, University of Iowa Carver College of Medicine, Iowa City, IA, USA

University of Iowa College of Public Health, Iowa City, IA, USA

Jennifer A. Sekeres, Pharm.D. Pharmacy Department, Cleveland Clinic, Cleveland, OH, USA

Audrey Wanger, Ph.D. Department of Pathology, University of Texas Medical School, Houston, TX, USA

Chapter 1
Antifungal Agents

Gerri S. Hall, Jennifer A. Sekeres, Elizabeth Neuner, and James O. Hall

Abstract This chapter lists the formulations for and primary uses of, and the mechanisms and spectrum of activity for the major classes of antifungal agents. Interpretation of laboratory results and comments about toxicity or adverse effects are noted where appropriate. When resistance has been found for the agents, this will be described with references to instances where it has been reported.

1.1 The Polyenes

1.1.1 Amphotericin

1.1.1.1 Brand Names and Formulations

Conventional amphotericin B (CAB; Fungizone™, others), IV formulation, lozenges, oral suspension
   Liposomal formulations:

Amphotericin B cholesteryl sulfate complex (ABCD; Amphotec)
Amphotericin B lipid complex (ABLC; Abelcet)
Liposomal amphotericin B (LAmB; AmBisome)

G.S. Hall, Ph.D. (✉)
Section of Clinical Microbiology, Department of Clinical Pathology,
Cleveland Clinic, Cleveland, OH 44195, USA
e-mail: hallg@ccf.org

J.A. Sekeres, Pharm.D. • E. Neuner, Pharm.D.
Pharmacy Department, Cleveland Clinic, Cleveland, OH, USA

J.O. Hall, D.P.M.
Staff Emeritus, Section of Podiatric Medicine, Cleveland Clinic, Cleveland, OH, USA

G.S. Hall (ed.), *Interactions of Yeasts, Moulds, and Antifungal Agents:*
How to Detect Resistance, DOI 10.1007/978-1-59745-134-5_1,
© Springer Science+Business Media, LLC 2012

1

1.1.1.2 Structure

Amphotericin is a polyene antifungal that consists of naturally derived products of *Streptomyces* spp. Amphotericin is lipophilic and insoluble in water; the conventional formulation is dispersed in sodium deoxycholate which is then reconstituted in 5% dextrose in water [53].

1.1.1.3 Primary Uses

Serious fungal infections caused by susceptible yeasts and moulds. It has indications for use for prophylaxis in patients with risk for serious fungal infections; for empiric therapy in patients who have neutropenia and fever; for treatment of candidiasis including candidal esophagitis and hepatosplenic candidiasis; for severe candidiasis including candidemia, endophthalmitis, and osteomyelitis; and for treatment of aspergillosis, cryptococcosis, and the dimorphic mycoses [53].

At one time, it was the first-line agent for invasive aspergillosis, but today voriconazole has replaced it in many cases. Amphotericin remains as second-line or salvage therapy for *Aspergillus* and other fungal infections. It remains as agent of choice for Zygomycetes infections, although newer azoles, such as posaconazole may have a greater role in this disease. The liposomal formulations have the same indications, although with the potential for lowered toxicity as compared to conventional amphotericin.

1.1.1.4 Spectrum of Activity

Amphotericin has a broad spectrum of activity against most yeasts, *Aspergillus* spp. and other hyaline and dematiaceous moulds and systemic dimorphic moulds to include:

Candida sp. with the possible exception of rare isolates of *C. lusitaniae* [4]
Cryptococcus neoformans
Aspergillus sp. (with exception of *A. terreus*) [7]
Other hyaline moulds, including most species of *Penicillium, Paecilomyces,* and *Scopulariopsis*
Dimorphic fungi to include *Histoplasma capsulatum, Blastomyces dermatitidis, Coccidioides immitis, and Paracoccidioides brasiliensis*

Note: *Pseudallescheria boydii* and *Scedosporium prolificans* are intrinsically resistant as are some species of *Trichosporon* spp. and *Fusarium* sp.

1.1.1.5 Mechanism of Action

The target of action is the plasma membrane ergosterol. These are the outcomes of its activity:

1. Amphotericin binds to ergosterol in cell wall membrane of fungi, resulting in alteration of membrane permeability by forming oligodendromes that function as pores through which there is a leakage of cellular contents. There is an efflux of potassium ions that then leads to disruption of the proton gradient and leakage of other intracellular molecules, causing death to the susceptible fungus [53].
2. Amphotericin causes an oxidation-dependent stimulation of macrophages either due to autoxidation of the drug in conjunction with formation of free radicals or due to the increase in membrane permeability [53].

Sterols, like cholesterol, in mammalian cells are also affected by amphotericin B. Amphotericin is a fungicidal agent. There is also some evidence that amphotericin acts as a proinflammatory agent and serves to stimulate innate host immunity. This process involves the interaction of amphotericin with Toll-like receptor 2 (TLR-2), the CD14 receptor, and stimulates release of cytokines, chemokines, and other immunologic mediators [53].

1.1.1.6 Pharmacokinetics: (Absorption/Distribution/Excretion)

There is poor oral absorption of amphotericin, and hence all systemic preparations are IV formulations.

The drug is distributed widely with higher concentrations in liver and spleen; lesser amounts in kidney and lung. It is, however, also distributed to adrenal glands, muscles, and other tissues in potentially therapeutic concentrations. Concentrations attained in pleural, peritoneal, and synovial fluids and in aqueous humor are reportedly about two thirds the concentration in plasma. Vitreous penetration of amphotericin is low (0–38%), and intraocular injections may be needed if amphotericin must be used [53]. Concentrations in CSF are approximately 3–5% of concurrent serum concentrations. To achieve fungistatic CSF concentrations, amphotericin B has been administered intrathecally, although this is a difficult way to administer the drug, and patient tolerability is often less than optimum. There are newer antifungal agents that more often used to treat fungal meningitis [40].

IV administration produces peak serum concentrations within 1 h and detectable levels may then persist for up to 24 h. Peak amphotericin B concentrations of 0.5–2 μg/ml are achieved with doses of 0.4–0.6 mg/kg.

Approximately 90% of the drug is protein-bound. Amphotericin is extensively metabolized and although the exact mechanism is poorly understood, it is presumed to be metabolized by the liver. Less than 10% unchanged drug is excreted in the urine, and bile or feces. No renal dose adjustment is necessary; amphotericin is not dialyzed. The elimination half-life is ~24 h.

All of the lipid formulations are >95% protein-bound as well (primarily to albumin) and also have long half-lives [53].

1.1.1.7 Pharmacodynamic Target

The pharmacodynamic target for predicting outcomes of amphotericin therapy is the peak serum level to mean MIC. A C_{max}/MIC ratio is suggested to be four for obtaining a 50% efficacy and ten for 100% maximal efficacy. (C = concentration). Drug levels are usually not needed during therapy and are infrequently measured. The drug demonstrates a concentration-dependent activity with a long post-antifungal effect [44].

1.1.1.8 Adverse Reactions/Drug Interactions

(a) Infusion-related events such as fever, chills, and myalgias are most common side effects and may occur in up to 50% of patients. These usually occur within the first hour of infusion. Some of these can be ameliorated if patients are premedicated with acetaminophen or diphenhydramine with or without meperidine. Slowing the infusions may also alleviate some of the infusion-related symptoms. Rarely, anaphylactoid reactions may occur and include hypotension, respiratory distress, hypoxia, and tachycardia [57].
(b) The most serious side effect of amphotericin is its propensity to produce nephrotoxicity in the form of azotemia, decreased glomerular filtration, and loss of the ability to concentrate urine. This is dose-limiting, and reversible renal function impairment occurs in 80% of patients. This side effect may present with renal tubular acidosis and hypokalemia and hyphomagnesemia [57].
(c) A reversible hypochromic, normocytic anemia can occur due to hemolysis or decreased erythropoietin production.
(d) Constitutional side effects to include nausea, vomiting, anorexia, weight loss, and a possible unpalatable "aftertaste" have also been reported to occur.

1.1.1.9 Drug Interactions

Amphotericin can increase the accumulation of renally cleared drugs such as 5-FC, fluconazole, ß-lactam antibiotics, and many other antimicrobials. The dose of amphotericin should be adjusted when given with any of these agents. The nephrotoxicity of amphotericin is enhanced when given along with aminoglycosides, cyclosporine, IV contrast dye, foscarnet, and many other agents. Minimizing coadministration of these latter drugs should be considered or use of the liposomal formulation of amphotericin.

Some of the side effects of amphotericin can be reduced with use of the lipid formulations of amphotericin. The lipid formulations of amphotericin are much less associated with the adverse side effects of conventional amphotericin.

Amphotericin B colloidal dispersion (ABCD, Amphocil, Amphotec) uses sodium cholesteryl as its lipid component. The bloodstream concentration is approximately equivalent at steady state to conventional amphotericin. Its safety is superior compared to amphotericin in regard to renal function, but has been however associated with dyspnea and hypoxia upon infusion [58].

Amphotericin B liquid complex (Abelcet) has as its lipid components dimyristoylphosphatidylcholine and dimyristoylphosphatidylglycerol. Its acute toxicity is eight to ten times less than conventional amphotericin in animal models; it results in equivalent bloodstream concentrations to conventional amphotericin B; its safety profile is superior to conventional amphotericin B in renal function and equivalent to conventional amphotericin B in infusional toxicity [58].

LAmB (AmBisome) has as its lipid component soy phosphatidylcholine, distearoyl phosphatidylglycerol, and cholesterol. In animal models, it has been shown to be 70–80 times less acutely toxic as compared to conventional amphotericin B. AmBisome yields higher bloodstream concentrations than conventional amphotericin B, and its safety profile is superior to conventional amphotericin B for renal function and infusion toxicity. It has been associated with back pain during infusion, however [57, 58]. Since the lipid preparations of amphotericin have a lower incidence of renal toxicity, if creatinine starts to rise during use of conventional amphotericin, switching to one of the lipid formulations has been shown to stabilize or improve renal function in a significant number of patients studied [53]. Alternatively, starting with liposomal formulations have become more standard of care if amphotericin is to be used for therapy.

1.1.1.10   Resistance

Primary or intrinsic resistance can be demonstrated by yeasts such as *C. lusitaniae* [4], *C. lipolytica*, and *C. guilliermondii* and *Trichosporon* spp.; secondary or developed resistance has been seen with some other species of *Candida* and rare isolates of *Cryptococcus neoformans*. Among the moulds, primary resistance is a characteristic of *Aspergillus terreus* [7], some species of *Fusarium*, and members of the *Scedosporium* genus (including *Pseudallescheria boydii*). The mechanism for resistance to amphotericin is an alteration in the cell membrane ergosterol that leads to a lowered affinity of the drug due to a lack of appropriate binding site [7, 53]. In addition, alterations in the content of the β-1,3-glucans in the cell wall of a fungus could increase stability of the wall and decrease the entrance of large molecules like the polyene antifungals. Amphotericin resistance in *A. terreus* has been studied, and the level of catalase was significantly higher than in *Aspergillus fumigatus*; since oxidative damage has been implicated in the action of amphotericin B, the high level of catalase may contribute to intrinsic resistance [7]. Secondary resistance can occur in fungi, although rarely, while a patient is on amphotericin.

1.1.1.11   MIC Interpretation

There are no interpretive criteria for amphotericin vs. yeasts or moulds, although an MIC of >1 µg/ml is often considered as indicative of yeast resistance [12, 38].

1.1.1.12   Comments

Because of the often severe kidney toxicity of this agent, amphotericin is not always the first antifungal agent considered in treating fungal infections. Some authors have suggested it is no longer the "gold standard for treatment of fungal infections [40]. There are at least three liposomal formulations of amphotericin B: Abelcet® (ABLC), Amphotec® (ABCD), and AmBisome® (LAmB) which afford the efficacy of amphotericin, with much less associated toxicity [57]. Mechanisms of action, mode of resistance development, and susceptibility testing are no different than for nonliposomal amphotericin.

1.1.2   Nystatin

1.1.2.1   Brand Name/Formulations

Mycostatin®; oral; cream and suspensions; also shampoos, ointment, and powder formulations, vaginal cream and suppository.

1.1.2.2   Primary Uses

Nystatin is used for the treatment of oral and vaginal candidiasis. Not for treatment of systemic fungal infections. Oral suspensions of Nystatin are usually cherry mint flavored and dosed as 100,000 USP/ml.

1.1.2.3   Spectrum of Activity

The activity of nystatin includes most yeasts, including *Candida albicans, C. parapsilosis, C. tropicalis, C. guilliermondii, C. krusei, and C. glabrata*; in vitro activity against dermatophytes including *Trichophyton rubrum* and *T. mentagrophytes* has also been demonstrated.

1.1.2.4 Mechanism of Action

Nystatin is a polyene antifungal agent, obtained from *Streptomyces noursei*. It can be either fungistatic or fungicidal to susceptible fungi. Nystatin binds to sterols in the cell membrane of yeasts, increasing the membrane permeability and consequent leakage of intracellular components. Nystatin has no activity against bacteria, protozoa, or viruses [44].

1.1.2.5 Pharmacokinetics: (Absorption/Distribution/Excretion)

GI absorption from oral doses is insignificant and most of the drug is excreted in the stool unchanged. In patients with renal insufficiency, receiving conventional dosages of oral therapy, significant plasma concentrations of nystatin may occasionally occur [44].

1.1.2.6 Adverse Reactions/Drug Interactions

Nystatin is usually well tolerated; however some oral irritation may occur; also, GI side effects including diarrhea, nausea, and vomiting may be present; rash and urticaria have been documented as well as rare cases of a Stevens-Johnson-like syndrome. Rare instances of tachycardia, bronchospasm, facial swelling, and nonspecific myalgias have been reported.

There are no reported drug interactions with nystatin.

It is listed as Category C in regard to any teratogenic side effects during pregnancy; it is unknown if the drug is excreted in human milk. Caution should be exercised when using during pregnancy or in nursing mothers.

1.1.2.7 Resistance

In vivo resistance has not been demonstrated, although *Candida* sp. other than *C. albicans* can become resistant in vitro upon repeated subcultures.

1.1.2.8 MIC Interpretations

Susceptibility testing is rarely if ever performed; CLSI guidelines for performing in vitro tests and their interpretation are not available.

1.1.2.9 Comments

Nystatin is not used for serious systemic fungal infections; however, a liposomal formulation is in Phase II and III trials and may afford another drug for systemic candidal infections, including against amphotericin B–resistant isolates, at least in vitro [3].

1.2 5-Fluorocytosine (5-FC)

1.2.1 Brand Name/Formulations

5-FC, Ancobon®

1.2.2 Structure

5-FC is a fluorinated pyrimidine analog of cytosine; it is deaminated to 5-fluorouracil after it enters the fungal cell, and this is further metabolized once in the cell.

1.2.3 Primary Uses

Never used along, but rather in combination with other antifungal agents, most commonly with amphotericin B for the treatment of cryptococcal meningitis; in addition, the combination has been shown to be active in vitro against *Candida* sp. 5-FC has also been recommended for use in synergy with other drugs for treatment of chromoblastomycosis.

1.2.4 Spectrum of Activity

The spectrum of activity of 5-FC includes most species of *Candida* (except for *C. krusei*) and *C. neoformans* and *Saccharomyces* spp. Dematiaceous moulds, responsible for chromoblastomycosis may also be susceptible, such as *Phialophora* and *Cladosporium*. *Aspergillus* spp., the Zygomycetes, and dermatophytes are all intrinsically resistant to 5-FC [53].

1.2.5 Mechanism of Action

5-FC is a fluorinated pyrimidine analog of cytosine that inhibits fungal protein synthesis by replacing uracil with 5-flurouracil in fungal RNA. Flucytosine also inhibits thymidylate synthetase via 5-fluorodeoxyuridine monophosphate (5-FUdRMP) and thus interferes with fungal DNA synthesis. Mammalian cells do not possess cytosine deaminase and hence cannot convert flucytosine to fluorouracil, so they are not usually affected adversely by this agent. 5-FC can be either fungicidal or fungistatic depending upon the species and strain of fungus [53].

1.2.6 Pharmacokinetics: (Absorption/Distribution/Excretion)

5-FC is well absorbed from the gastrointestinal tract, with greater than 80–90% absorption following oral dose. Peak serum levels occur 1–2 h after ingestion.

The drug is well distributed, with a volume of distribution of 0.6–0.9. Bone, peritoneal fluid, and synovial fluid have 5-FC levels several folds higher than concurrent serum levels [53]. 5-FC achieves good CSF levels (75% of plasma level).

5-FC is minimally protein-bound (<5%). 5-FC is minimally metabolized by the liver and >90% of the drug is excreted renally, in urine, as unchanged drug. A small portion of dose is found excreted in feces. 5-FC has a very short post-antifungal effect [44].

The half-life in patients with normal renal function is 2.5–5 h, but this can be prolonged to as much as 250 h in patients with moderate to severe azotemia. Renal dose adjustment is necessary. 5-FC is dialyzed significantly [57].

1.2.7 Pharmacodynamic Target

The pharmacodynamic target is Time (T) above the MIC ($T>$MIC); this should be at least 25%.

1.2.8 Adverse Reactions/Drug Interactions

5-FC has very low toxicity associated with its use:

(a) Mainly GI symptoms, such as nausea, abdominal pain, and diarrhea; elevated levels of hepatic transaminases may occur.

(b) Rare ulcerations and enterocolitis have been reported.
(c) Some rash, ataxia, paresthesias, and confusion may occur but rarely.
(d) Bone marrow depression can occur with use of 5-FC resulting in neutropenia, anemia, and thrombocytopenia. The suppressive effects are usually seen when blood levels exceed 100–125 µg/ml [53].
(e) In patients with impaired renal function, monitoring of creatinine levels is usually suggested because high creatinine levels can elevate flucytosine levels. There are increased risks for toxicity with higher drug levels.
(f) 5-FC does cross the placenta and is teratogenic; it should not be administered during pregnancy [53].

1.2.9   Resistance

Resistance to 5-FC develops quickly in most fungi especially if used alone. Primary resistance is seen in some strains of *Candida* sp., like *C. krusei*. Resistance develops due to mutations in cytosine permease or cytosine deaminase, which are the enzymes through which the drug usually enters the cell and is converted to 5-fluorouracil respectively [41].

1.2.10   MIC Interpretations

MICs have been developed by CLSI for yeast as follows:
$\leq$4 µg/ml = S; 8–16 µg/ml = I; $\geq$32 µg/ml = R [12].

1.2.11   Comments

5-FC should never be used alone since resistance develops quickly. Synergy with amphotericin has been documented for cryptococcal meningitis.

1.3   The Azoles

The azoles are a class of antifungal agents that inhibit the cytochrome P450–dependent enzyme lanosterol C$_{14}$ α-demethylase. This in general causes a disruption of membrane synthesis, a depletion of ergosterol which then leads to an increase in toxic methylated sterol precursors in the cell membrane. The overall effect is felt to possibly result in fungicidal activity by many of the azoles due to an increased sensitivity of the susceptible fungus to oxygen-dependent microbicidal systems of the host. The area under the curve (AUC) to MIC ratio is the primary predictor of drug efficacy.

Azoles differ in their affinity for the 14-alpha-demethylase enzyme, and this difference is responsible for their varying antifungal potency and varying spectrum of activity. There are differences among the older and newer azoles and that will be discussed for each agent individually. Availability of oral and IV formulations will vary with each specific agent.

Toxic side effects are usually low, but drug interactions can occur with the azoles. Some of the most significant drug interactions of triazoles are drug elevations of cyclosporine, tacrolimus, and sirolimus, most calcium channel blockers, most benzodiazepines, many statins and steroids, warfarin, and rifabutin. Carbamazepine, phenobarbital, phenytoin, rifampin, and rifabutin significantly decrease azole concentrations. Increased blood levels of terfenadine, astemizole, cisapride, pimozide, and quinine can cause QTc prolongation and predispose to torsades de pointes. In addition, increased cytotoxin chemotherapy-related toxicity can be caused by concomitant treatment with triazoles (except for fluconazole) and vinca alkaloids, cyclophosphamide, vinorelbine, and busulfan. The absence of a known interaction does not mean that it cannot occur especially with the newer azoles [60].

There are two classes of azoles, the Imidazoles (possessing two nitrogen atoms in the azole ring) that include mainly the topical agents clotrimazole, miconazole, and ketoconazole. The other groups are the triazoles (possessing three nitrogen atoms in the azole ring) and include fluconazole, itraconazole, voriconazole, and posaconazole. The newer azoles like voriconazole and posaconazole have been developed to overcome the limited efficacy of fluconazole against *Aspergillus* spp. and other moulds and to improve the absorption, tolerability, and drug interaction profile of itraconazole [60].

The triazoles do vary in their inhibition of CYP450 enzymes. Fluconazole, voriconazole, and posaconazole have a moderate inhibition on CYP3A4, whereas the inhibition by itraconazole is strong. Itraconazole and posaconazole have no inhibitory effects on CYP2C9 or CYP2C19. Fluconazole has a strong inhibitory effect on CYP2C9 and moderate effect on CYP2C19. Voriconazole has a moderate inhibitory effect on CYP2C9 and weak inhibitory effect on CYP2C19 [60].

1.3.1 The Imidazoles

1.3.1.1 Clotrimazole

Brand Name/Formulations

Lotrimin®, Gnye-Lotrimin®, Mycelex®, Mycelex-G®; topical in form of creams or lotions. Lotrisone® cream and lotion contain combinations of clotrimazole, a synthetic antifungal agent, and betamethasone dipropionate, a synthetic corticosteroid, for dermatologic use. No systemic formulations of clotrimazole are available.

Primary Uses

Cutaneous candidiasis; *Candida* infections of mucous membranes and mucocutaneous junctions, such as perianal, intertriginous (between digits), and vaginal areas. Mycelex® Troches are administered only as a lozenge that must be slowly dissolved in the mouth. Mycelex® Troches are indicated prophylactically to reduce the incidence of oropharyngeal candidiasis in patients immunocompromised by conditions that include chemotherapy, radiotherapy, or steroid therapy utilized in the treatment of leukemia, solid tumors, or renal transplantation.

Spectrum of Activity

In vitro, clotrimazole demonstrates fungistatic activity at concentrations of drug up to 20 µg/ml and may be fungicidal in vitro against *C. albicans* and other species of the genus *Candida* at higher concentrations; it is active in vitro against the dermatophytes *Trichophyton mentagrophytes, T. rubrum, Epidermophyton floccosum*, and *Microsporim canis* [46].

Mechanism of Action

Clotrimazole is an imidazole which causes changes in the fungal cell membrane that result in leakage of intracellular compounds outside of the susceptible cell; clotrimazole may also act to interfere with amino acid transport into fungus [18].

Pharmacokinetics: (Absorption/Distribution/Excretion)

When administered topically only, there is no distribution to other tissues. It is well absorbed in humans following oral administration and is eliminated as inactive metabolites. After oral administration, peak serum concentrations are 1.16–1.29 µg/ ml within 1 h and drop quickly thereafter [9].

Adverse Reactions/Drug Interactions

Contraindicated for use if there is a hypersensitivity to clotrimazole or any of the components of the formulations. There are rare effects of erythema, urticaria, pruritus, stinging, and possible blistering. GI disturbances and alterations in hepatic and adrenal functions have been reported [44].

Category C in pregnancy. There are no adequate and well-controlled studies in pregnant women. Clotrimazole troches should be used during pregnancy only if the potential benefit justifies the potential risk to the fetus [44].

Resistance

Natural resistance to clotrimazole is rare in the fungi; there has been a strain of *Candida guilliermondii* reported to have primary clotrimazole resistance. No single-step or multiple-step resistance to clotrimazole has developed during successive passages of *Candida albicans* in the laboratory; however, individual organism tolerance has been observed during successive passages in the laboratory. Such in vitro tolerance has resolved once the organism has been removed from the antifungal environment. In one study of vaginitis, cross-resistance was determined between fluconazole and over-the-counter antifungal agents like clotrimazole. Spontaneous resistant mutants to clotrimazole were found to be cross resistant to fluconazole in vitro [14].

MIC Interpretive Criteria

There is no standardized method for in vitro susceptibility testing. In a study comparing in vitro activity of azoles vs. *Candida* spp. involved in vulvovaginitis, >94.3% were found susceptible to clotrimazole at MIC </=1 µg/ml [46].

Comments

Used only for cutaneous and mucous membrane candidal infections. No indications for use of troches for treatment of systemic candidiasis. Vaginal cream not to be used orally or systemically.

1.3.1.2 Miconazole

Brand Name and/or Formulations

Monistat-Derm (miconazole nitrate 2%) cream contains miconazole nitrate* 2%, formulated into a water-miscible base consisting of pegoxol 7 stearate, peglicol 5 oleate, mineral oil, benzoic acid, butylated hydroxyanisole, and purified water.

Primary Uses

For topical application in the treatment of tinea pedis, tinea cruris, and tinea corporis caused by *Trichophyton rubrum, Trichophyton mentagrophytes*, and *Epidermophyton floccosum*, in the treatment of cutaneous candidiasis, including vaginal and oral infections, and in the treatment of tinea versicolor.

Spectrum of Activity

Miconazole nitrate inhibits the growth of the common dermatophytes, *Trichophyton rubrum, Trichophyton mentagrophytes*, and *Epidermophyton floccosum*; inhibits *Candida albicans*, and inhibits *Malassezia furfur*, the agent of tinea versicolor. Miconazole has in vitro activity against dimorphic fungi and hyaline moulds [18].

Mechanism of Action

Miconazole is an imidazole; inhibits cytochrome P450-dependent enzyme lanosterol C $_{14}$ α-demethylase. This leads to disruption of membrane synthesis and depletion of ergosterol that causes an increase in toxic methylated sterol precursors in the membrane. The action of most azoles is considered fungistatic; however, recent studies have indicated a fungicidal activity of micronazole specifically due to accumulation of drug-induced reactive oxygen species within the fungal organism that results in oxidative damage and cell death [5].

Pharmacokinetics: (Absorption/Distribution/Excretion)

Oral absorption is 25–30%. Topical miconazole is not systemically absorbed (<0.1%), nor is the vaginal dose (1.4%). Oral miconazole is eliminated in the feces as unchanged drug [57].

Adverse Reactions/Drug Interactions

There have been isolated reports of irritation, burning, maceration, and allergic contact dermatitis associated with the application of Monistat-Derm.

Resistance

Cross-resistance with other azoles can occur, especially in strains of *C. glabrata*. In more recent studies, clinical isolates of *Candida* spp. have been shown to be susceptible to miconazole even in isolates that have developed fluconazole resistance and when isolates are exposed to increasing concentrations of miconazole in vitro [19].

MIC Interpretations

There are no breakpoints vs. *Candida* spp. for miconazole, and susceptibility testing for this agent is usually not performed in clinical laboratories because of lack of its use clinically in face of newer triazoles. Primary use of miconazole is for dermatophyte and *Candida* superficial skin infections.

Comments

Rapid intravenous administration of miconazole may lead to life-threatening cardiac arrythmias. There is little use of miconazole outside of treatment of dermatophyte infections, cutaneous *Candida* infections, and tinea versicolor. With the advent of new azoles and echinocandins that offer more broad-spectrum coverage with lesser toxicity, use of miconazole is limited. There are, however, more recent studies in the literature advocating miconazole for first-line treatment of oropharyngeal candidiasis [24].

1.3.1.3 Ketoconazole

Brand Name/Formulations

Nizoral® 2% cream, tablet, and shampoo available in USA.

Primary Uses

Used for the topical treatment of tinea corporis, tinea cruris, and tinea pedis; treatment of tinea versicolor; treatment of cutaneous candidiasis and seborrheic dermatitis.

Spectrum of Activity

Dermatophyte infections caused by *Trichophyton rubrum, T. mentagrophytes*, and *Epidermophyton floccosum*; *Candida* spp.; *Malassezia furfur*. Ketoconazole does have in vitro activity against dimorphic fungi and dematiaceous agents of chromoblastomycosis [18].

Mechanism of Action

Ketoconazole is an imidazole; inhibits cytochrome P450-dependent enzyme lanosterol C $_{14}$ α-demethylase. This leads to disruption of membrane synthesis and depletion of ergosterol that causes an increase in toxic methylated sterol precursors in the membrane. The action is often fungicidal, increasing the sensitivity of the fungus to oxygen-dependent microbicidal systems of the host.

Pharmacokinetics: (Absorption/Distribution/Excretion)

Peak plasma concentrations are achieved 1–2 h after oral administration. It requires acidity for dissolution and absorption; if patient is receiving antacids, anticholinergics, H$_2$ blockers, they should be given at least 2 h after administration of the ketoconazole. There is a need to acidify the tablet of ketoconazole if given to achlorhydric patients. It is highly protein-bound (99%); although there is distribution to body fluids and tissues, there is only negligible amount in the CSF. About 13% is excreted via urine; majority is excreted via bile [44].

Ketoconazole demonstrates time-dependent pharmacokinetics and has a long post-antifungal effect.

The pharmacodynamic target for efficacy is the AUC (area under the curve)/MIC of 25.

Adverse Reactions/Drug Interactions [44]

(a) The major toxicity is that of hepatic toxicity, with an incidence of 1:10,000, and this is usually reversible. Obtaining baseline levels of liver function tests and monitoring these through treatment is important44.
(b) There have been cases of hypersensitivity in the form of urticaria, lowered serum testosterone levels, and decreased ACTH-induced corticosteroid levels with administration of ketoconazole.
(c) There are no known controlled studies in the pregnant patient; teratogenic effects have been seen in animals given large doses of ketoconazole.
(d) Cardiac complications have been found with coadministration of terfenadine and astemizole and cisapride.
(e) The package contains contraindications for use of ketoconazole along with these agents.
(f) Ketoconazole can increase blood levels of cyclosporine, theophylline, and anticoagulants.

Resistance

Cross-resistance with other azoles (fluconazole) occurs in *Candida glabrata* clinical isolates [47].

MIC Interpretations

There are breakpoints vs. *Candida* spp. for ketoconazole, although susceptibility testing for this agent is usually not performed in clinical laboratories because of lack of its use due to newer, more efficacious and less toxic triazoles [12]. Primary use of ketoconazole is for dermatophyte and *Candida* superficial skin infections.

Comments

There is little use of ketoconazole outside of treatment of dermatophyte infections, cutaneous *Candida* infections, and tinea versicolor. With the advent of new azoles and echinocandins that offer broader spectrum coverage with lesser toxicity, use of ketoconazole is limited.

1.3.2   Triazoles

1.3.2.1   Fluconazole

Brand Names/Formulations

Diflucan®; tablets, oral suspension, and IV formulations available

Primary Uses

Used for vaginal candidiasis, oropharyngeal and esophageal candidiasis, cryptococcal meningitis; prophylaxis in bone marrow transplant patients to decrease development of disseminated candidiasis. Fluconazole has also been used for infections of *Coccidioides immitis* [53].

Spectrum of Activity

Candida sp., to include *C. albicans*, most strains of *C. tropicalis*, and *C. parapsilosis*. *C. krusei* is intrinsically resistant; most strains of *C. glabrata* demonstrate reduced susceptibility. *C. neoformans* is susceptible; *Rhodotorula* spp. and *Saccharomyces* spp. are usually susceptible [53].
Note: *Trichosporon* spp. are usually considered resistant. There is also no activity against moulds such as *Aspergillus* spp., *Fusarium* spp., *Pseudallescheria boydii*, and the Zygomycetes.

Mechanism of Action

Fluconazole inhibits cytochrome P450-dependent enzyme lanosterol C_{14} α-demethylase. This leads to disruption of membrane synthesis and depletion of ergosterol that causes an increase in toxic methylated sterol precursors in the membrane. The action is often fungicidal, increasing the sensitivity of the fungus to oxygen-dependent microbicidal systems of the host.

Pharmacokinetics: (Absorption/Distribution/Excretion)

Fluconazole is water soluble and is absorbed very well from the GI tract. It is not affected by food or gastric pH. Fluconazole is distributed widely in the body, including the cerebrospinal fluid (70–80% serum levels). Vitreal penetration is 28–75% of serum levels [8, 46]. Adequate levels can be achieved in saliva and sputum; lower levels in vaginal fluid. Oral absorption remains unchanged in patients receiving acid-suppressive therapy (protein pump inhibitors, H_2 blockers) [53].

Bioavailability is >90%; peak serum concentrations are achieved in 1–2 h after a usual dose. Fluconazole is ~12% protein-bound. Half-life of fluconazole is 24–30 h; steady state following oral administration is achieved after 5–7 days.

Renal excretion is the usual mode of elimination of the drug (80% is excreted unchanged in urine). Hepatic CYP2C9 enzyme plays a minor role in metabolism of fluconazole. Renal dose adjustment is needed [20, 60]. Patients with a reduced creatinine clearance should receive lesser doses (as much as 50% less); 100% of the regular dose is recommended as a loading dose after hemodialysis [60].

Fluconazole, as with other azoles, has a time-dependent kinetics; the post-antifungal effect is long.

Pharmacodynamic Target

The pharmacodynamic target for efficacy is to achieve an AUC (area under the curve)/MIC of ≥25.

Adverse Reactions/Drug Interactions

In general, fluconazole is well tolerated, and serious adverse side effects are rare. Headache, nausea, vomiting, and abdominal pain and diarrhea are the more common, although rare, side effects [60]. Fluconazole is contraindicated if the patient is hypersensitive to an "azole."
Fluconazole:

(a) Should not be coadministered with terfenadine (antihistamine) or cisapride (for gastric reflux). There are reports of cardiac complications when used with cisapride in particular.

(b) May have decreased serum concentration if administered with antacids, H_2 receptor antagonists, proton pump inhibitors, sucralfate, and diadenosine; fluconazole increases the concentration of INH, rifampin, phenytoin, carbamazepine,and phenobarbital.

(c) May increase the concentration of other drugs when given with fluconazole: anticoagulants, oral hypoglycemics, cyclosporine, tacrolimus, some protease inhibitors, INH, rifampin, and cyclophosphamide.

Rare side effects include:

(a) Fluconazole has been associated with rare cases of hepatic toxicity.

(b) Few reports of rashes and exfoliative skin disorders; rare cases of alopecia have occurred, but these are reversible and dose-dependent.

(c) Some associated gastrointestinal distress has been seen.

(d) Headache and transient increases in LFTs.

Fluconazole is a Category C drug, and its use in pregnancy should be avoided [60].

Resistance

Primary intrinsic resistance in *C. krusei*; decreased susceptibility or resistance is common *in C. glabrata*; secondary resistance may develop in any species of *Candida* and in patients on therapy for *C. neoformans*.

Resistance to azoles is due to decreased membrane permeability, presence of multidrug efflux transporters, or altered or overproduction of the target enzyme, which is encoded on the ERG11 gene [6, 47].

MIC Interpretations

Interpretive guidelines are available for *Candida* sp. and are as follows:
≤8 µg/ml = S; 16–32 µg/ml = S-DD (susceptible dose-dependent); ≥64 µg/ml = R [12].

Comments

Fluconazole is most often considered a yeast and not a mould agent. Most *Candida* sp. (except *C. krusei* and *C. glabrata*) and *Cryptococcus neoformans* are fully susceptible.

1.3.2.2 Itraconazole

Brand Names/Formulations

Sporanox®; available in capsules and an oral solution. *The IV formulation is no longer available.* Itraconazole is solubilized in the oral solution by the pharmacologically inactive oligosaccharide, hydroxypropyl-β-cyclodextrin. This vehicle is released when the itraconazole molecule makes contact in the lipid membrane of the enteric cell [60].

Primary Uses

Pulmonary and extrapulmonary blastomycosis; most forms of histoplasmosis; pulmonary and extrapulmonary aspergillosis in patients who are intolerant to or refractory to amphotericin B therapy; used for treatment of onychomycosis caused by dermatophytes. It has been relegated to salvage therapy for invasive aspergillosis in light of newer azoles that are available. It is approved for treatment of allergic bronchopulmonary aspergillosis.

Spectrum of Activity

Candida albicans and other *Candida* sp. except *C. krusei* which is intrinsically resistant. Active against *Aspergillus* sp. *C. neoformans*, the dimorphic systemic fungi including *Histoplasma capsulatum, Blastomyces dermatitidis, Coccidioides immitis, Paracoccidioides brasiliensis.* Itraconazole has activity against *Sporothrix schenckii*, the dermatophytes (*Microsporum, Trichophyton, Epidermophyton*), and many of the dematiaceous moulds [30].

Mechanism of Action

Itraconazole inhibits cytochrome P450-dependent enzyme lanosterol C_{14} α-demethylase. This leads to disruption of membrane synthesis, depletion of ergosterol that causes an increase in toxic methylated sterol precursors in the membrane. The action is often fungicidal, increasing the sensitivity of the fungus to oxygen-dependent microbicidal systems of the host.

Pharmacokinetics: (Absorption/Distribution/Excretion)

Itraconazole is water insoluble, and its bioavailability is variable after oral ingestion. The oral solution is better absorbed in the fasting state and generally achieves higher bioavailability [53, 60]. Itraconazole requires an acidic environment for optimum absorption; peak serum concentrations occur in 2.5–4 h, depending upon formulation of drug and whether it is taken with food. Itraconazole is highly lipophilic. It is highly protein-bound, with <1% available as free drug. It is extensively distributed to tissues, especially fatty tissues, like omentum, liver, and kidneys. Skin, lung, and female GU tract have increased levels above that in plasma, but negligible concentrations are achieved in CSF, aqueous fluid, and saliva [20, 31].

Primarily, hepatic metabolism; biliary excretion is only about 3–18%. The major liver metabolite is hydroxyitraconazole, and it does possess antifungal activity similar to that of the parent drug. No renal dose adjustment is needed. The half-life of itraconazole is 24–30 h, but is prolonged in patients who have hepatic dysfunction, and in these patients, drug dose adjustments, liver function testing, and drug interactions need to be assessed [20, 53, 60].

Itraconazole follows time-dependent pharmacokinetics as with other triazoles and has a long post-antifungal effect.

Pharmacodynamic Target

The pharmacodynamic target for efficacy is the AUC (area under the curve)/MIC of ≥25.

Adverse Reactions/Drug Interactions

Interactions can occur with other agents that are metabolized by P450 enzyme systems, that is, terfenadine, cisapride, triazolam, astemizole, or midazolam. Antacids, anticholinergics, antispasmodics, sucralfate or anything that increases GI pH may decrease absorption. Itraconazole usually causes an increase in their concentration [60].

Itraconazole and voriconazole are inhibitors of gastric P-glycoprotein, a transmembrane efflux pump that limits exposure to many drugs by inhibiting GI absorption. The inhibition of P-glycoprotein by itraconazole and posaconazole may lead to increased system exposure of drugs affected by this transport system. Interactions caused by itraconazole have been extensively studied; drugs with known interactions to itraconazole should be used with caution [60].

Although itraconazole is usually well tolerated, adverse reactions can include [60]:

(a) Gastrointestinal distress (nausea and vomiting which are dose-dependent), rash, headache and dizziness, pedal edema, and increases in LFTs.
(b) Congestive heart failure and pulmonary edema have been reported with itraconazole capsules, and these should not be administered for onychomycosis, for example, in patients with evidence of ventricular dysfunction.
(c) Use in any patient with liver dysfunction should be discouraged or monitored regularly. Serious hepatotoxicity can lead to liver failure, although this is rare.
(d) The presence of cyclodextrin in the oral solution can cause diarrhea [60].

Itraconazole is a Category C drug and should not be used in pregnancy and is contraindicated during lactation [60].

Resistance

Resistance is often demonstrated by *C. krusei, Saccharomyces* sp., and *Rhodotorula* sp. Zygomycetes and *Fusarium* sp. are often resistant. *Pseudallescheria boydii* may be resistant (about 50% of strains in vitro). Resistance in *Aspergillus* spp. has been increasing [56]. Resistance to itraconazole, like other azoles, is due to decreased membrane permeability; presence of multidrug efflux transporters; or altered or overproduction of the target enzyme, which is encoded on the ERG11 gene [47].

MIC Interpretation

Interpretive criteria available in the CLSI document for itraconazole vs. *Candida* sp. are: $\leq$0.125 µg/ml = S; 0.25–0.5 µg/ml = S-DD (susceptible dose-dependent); $\geq$1 µg/ml = R [12].

Comment

Liver enzymes are usually monitored periodically especially in any patient that is receiving itraconazole for longer than a month. Itraconazole is often considered to be a mould rather than a yeast agent.

1.3.2.3 Voriconazole

Brand Name/Formulations

Vfend™; It is available in both intravenous and oral formulations. It is a low molecular weight poorly water soluble triazole with a chemical structure similar to fluconazole. Solubility of the intravenous formulation is achieved by sulfobutyl ether β-cyclodextrin (SBECD), a molecule similar to that used for solubilizing itraconazole (hydroxypropyl-β-cyclodextrin). There is no SBECD in the oral formulation of voriconazole [60].

Primary Uses

FDA-approved indications as of 2005 for voriconazole are for primary treatment of aspergillosis; salvage therapy for serious fungal infections due to *Fusarium* sp., and *Scedosporium* sp. (*P. boydii*) in patients refractory to or intolerant to other therapy;

candidemia in nonneutropenic patients; disseminated candidal skin infection or
Candida infections of abdomen, kidney, and bladder wall; esophageal candidiasis.

Spectrum of Activity

Candida spp., including *C. krusei* and *C. glabrata* that are resistant to fluconazole.
In general, the MICs of voriconazole for *C. albicans* are 1–2 log lower than the
MICs for fluconazole; however, some highly fluconazole-resistant strains of
C. glabrata may demonstrate higher MICs to voriconazole as well [26]. Other
susceptible yeast include *Cryptococcus neoformans* and *Trichosporon beigelii* and
Saccharomyces cerevisiae [26]. Its spectrum of action includes the dimorphic
systemic fungi including *Histoplasma capsulatum, Blastomyces dermatitidis,
Coccidioides immitis,* and *Paracoccidioides brasiliensis.* Voriconazole is not as
active against *Sporothrix schenckii.*

Moulds including *Aspergillus* spp., *Fusarium* sp., *Paecilomyces* spp., *Bipolaris*
sp., *Alternaria* sp., *P. marneffei,* and some strains of *P. boydii* are also susceptible
in vitro [39].
Note: Zygomycetes are not susceptible to voriconazole [26, 53].

Mechanism of Action

Voriconazole inhibits cytochrome P450-dependent enzyme lanosterol C_{14}
α-demethylase. This leads to disruption of membrane synthesis, depletion of
ergosterol that causes an increase in toxic methylated sterol precursors in the
membrane. The newer azoles, like voriconazole also inhibit 24-methylene dihydro-
lanosterol demethylase to give increased activity against moulds. The action is often
fungicidal vs. moulds, increasing the sensitivity of the fungus to oxygen-dependent
microbicidal systems of the host. In vitro, data suggests static rather than cidal
activity against some yeasts.

Pharmacokinetics: (Absorption/Distribution/Excretion)

Voriconazole, like fluconazole, has excellent absorption and is >95% bioavailable
especially when given on an empty stomach; maximal plasma levels achieved after
1–2 h whether administered orally or via IV; no effects on absorption of voriconazole when administered with agents that increase the gastric pH. Fatty foods have
been found to reduce bioavailability of the oral preparations by 80%. Voriconazole
is extensively distributed into tissues (volume of distribution is 4.6 l/kg [1]), with
good levels in brain, CSF (level is ~46% of serum levels), and the eye (noninflamed
vitreous level is 53% and aqueous is 38%) [60]. Metabolism is in the liver, mainly
by CYP2C19 and CYP3A4 and to a lesser extent by CYP2C8/9. Excretion does not
depend on renal function [1]. The primary route of excretion is hepatic, and there

are specific dosing recommendations for hepatic impairment (reductions in dose of up to 50% is recommended) [1]; voriconazole should be used with caution in patients with creatinine clearance of <50 ml/min [60]. Only about 2% of the active drug is found in urine [60].

Voriconazole is estimated to be 58% bound to serum proteins [1, 20].

Steady state following oral administration is achieved after 5–6 days with oral or IV administration, but steady-state concentrations are reached within 24 h with a loading dose. Peak plasma concentrations are in the range of 3–6 µg/ml, but can vary considerably between patients [1, 20].

Voriconazole, like other azoles, has time-dependent kinetics; the post-antifungal effect is long. The pharmacokinetics is nonlinear in adults due to saturable metabolism, wide intersubject variability, and metabolism via CYP2C19 which exhibits genetic polymorphism. Up to 20% of non-Indian Asians have low CYP2C19 activity, and voriconazole serum levels are up to four times higher than those found in Caucasian and African American populations in which the poor metabolizer status is uncommon. The unpredictability of patient's enzymatic activity has generated an increased interest in the routine use of voriconazole serum level determinations [1, 53].

Pharmacodynamic Target

The pharmacodynamic target for efficacy is to achieve an AUC (area under the curve)/MIC of ≥25.

Adverse Reactions/Drug Interactions

Voriconazole is well tolerated; however, there are some unique side effects including visual disturbance, consisting of blurred vision, altered color discrimination, or photophobia, which are transient, reversible, and dose-dependent. Mild transaminase elevations have been reported in 10% of patients, but is reversible [1]. Patients should be warned to avoid extended exposure to sunlight during voriconazole treatment. Patients with lactose intolerance should not be given oral formulations of voriconazole since the vehicle it is given with contains lactose [1].

Due to accumulation of the IV voriconazole vehicle SBECD, the drug should be used with caution in patients with creatinine clearance of <50 ml/min [60].

Voriconazole has extensive interactions with other agents that are metabolized by P450 enzyme systems, that is, terfenadine, cisapride, triazolam, astemizole, or midazolam. Voriconazole is a substrate and inhibitor of CYP2C19, 2C9, and 3A4 and has a similar drug interaction profile to itraconazole. Agents metabolized via these pathways are likely to have interactions and require potential dosage adjustments with voriconazole. Voriconazole levels can be affected by simultaneous use of drugs such as rifampin, rifabutin, macrolide antibiotics, phenytoin, and omeprazole [53].

Adverse effects include:

(a) Ocular effects to include increased brightness, blurred vision, altered visual perception, photophobia, altered color perception, and ocular discomfort. These side effects can be seen in up to 30% of individuals that are on voriconazole, although these do not always cause a discontinuation of the drug. These are most often seen quickly after the dose is given (within 30 min) and during the first week of therapy.

(b) Cardiac effects including prolonged QTc interval, and rare causes of torsades de pointes reported.

(c) There can be increases in liver function tests especially related to high plasma concentrations and longer duration of therapy (≥ 7 days). Monitoring for hepatic toxicity is recommended prior to therapy, within 2 weeks of its initiation and every 2–4 weeks during therapy. The safety of using voriconazole in patients with severe liver disease is uncertain [1, 53].

(d) Rash has been noted in up to ~19% patients in clinical trials with voriconazole. Most are mild to moderate occurrences; however, the package insert does caution about severe rash, including Stevens-Johnson.

(e) Kidney toxicity in patients with renal impairment.

(f) Poor metabolizers of P450 cytochrome system will have four times the level of voriconazole of others that are good metabolizers.

(g) Infusion-related side effects include anaphylactoid like reactions immediately upon initiating infusion.

(h) Visual and rarely auditory hallucinations can occur in less than 5% of patients. This occurs with IV formulations and disappears with continued treatment [60].

Voriconazole is teratogenic in animals and is considered a Category D drug that should not be used in pregnancy, nor in woman who are breastfeeding [1, 60].

Resistance

Resistance to azoles is due to decreased membrane permeability, presence of multidrug efflux transporters, or altered or overproduction of the target enzyme, which is encoded on the ERG11 gene. Resistance may develop while on voriconazole. Yeast with increased fluconazole MICs may also have increased voriconazole MICs [41, 47].

MIC Interpretations

CLSI-defined interpretive breakpoints for voriconazole are ≤ 1 μg/ml = S; 2.0 μg/ml = S-DD (susceptible dose-dependent); ≥ 4 μg/ml = R [12].

Comments

There have been some reports of increased incidence of infections with Zygomycetes in association with increased use of voriconazole [29, 39].

1.3.2.4 Posaconazole

Brand Name/Formulations

Noxafil; cherry-flavored oral suspension using polysorbate 80 as the emulsifying agent; posaconazole is insoluble in water so there is no IV formulation available. Posaconazole is a lipophilic triazole with structure similar to that of itraconazole.

Primary Uses

Posaconazole was approved by FDA initially for the prophylaxis of invasive *Aspergillus* and *Candida* sp. infections in high risk, severely immunocompromised patients ≥13 years old, including hematopoietic stem cell transplant recipients with graft vs. host disease and patients with hematologic malignancies with prolonged neutropenia with chemotherapy [23, 36]. An oral suspension was in addition approved (October 2006) for use in the treatment of oropharyngeal candidiasis, including cases that are intolerant or refractory to itraconazole and/or fluconazole therapy [23]. Among the azoles, posaconazole has a significant role in treatment of Zygomycetes.

Spectrum of Activity

Posaconazole has a broad spectrum of activity, vs. yeasts, filamentous, and dimorphic fungi. This includes *Candida* sp., *Cryptococcus neoformans, Aspergillus* sp., *Rhizopus* sp., *Blastomyces dermatitidis, Histoplasma capsulatum, Coccidioides immitis*, dermatophytes, and dematiaceous fungi. Posaconazole has not been given an indication for salvage therapy of refractory fungal infections, but there are reported uncontrolled studies with good response rate when used for zygomycosis, histoplasmosis, fusariosis, coccidioidomycosis, or in patients with chronic granulomatous disease [23, 60].

Mechanism of Action

Posaconazole is a triazole; it inhibits cytochrome P450-dependent enzyme lanosterol C_{14} α-demethylase. This leads to disruption of membrane synthesis, depletion of ergosterol that causes an increase in toxic methylated sterol precursors in the membrane. The end result is an abnormality in the fungal membrane permeability and a lack of coordinated chitin synthesis [1, 36]. The action of posaconazole is often fungicidal vs. moulds, increasing the sensitivity of the fungus to oxygen-dependent microbicidal systems of the host.

Pharmacokinetics: (Absorption/Distribution/Excretion)

Oral administration should be done with food or a liquid nutritional supplement to enhance absorption. Posaconazole has a large volume of distribution [1]. The absorption of posaconazole is not affected by changes in gastric acidity, but its bioavailability can be increased up to 400% if taken with fatty meals [1]. Peak serum concentrations have shown considerable interpatient variability for as yet undetermined reasons. Therapeutic drug monitoring may be considered in some cases [53].

Posaconazole is metabolized by the liver and undergoes minimal glucuronidation and is converted to biologically inactive metabolites [1]. Seventy-seven percent is excreted unchanged in the feces and 14% in urine as multiple glycuronidated products [1, 53]. A small amount is excreted unchanged in the urine. Posaconazole is >98% protein-bound and has a long half-life of 16–35 h [1, 21, 53].

Posaconazole like other azoles has dose-dependent kinetics, and a saturable absorption greater than 800 mg/day, thus oral loading is not possible and steady state is not achieved until after 7–10 days of therapy; the post-antifungal effect is long [20]. Although not metabolized through liver CYP450 system enzymes, posaconazole is a moderate inhibitor of CYP3A4, and there will be increased blood levels of drugs that use the CYP3A4 metabolic pathway when they are coadministered with posaconazole. Posaconazole is not cleared with hemodialysis, and thus there is no need for dose adjustment in patients with moderate to severe renal disease [20, 60].

Pharmacodynamic Target

The pharmacodynamic target for efficacy is to achieve an AUC (area under the curve)/MIC of ≥25.

Adverse Reactions/Drug Interactions

In a study of 428 patients with posaconazole treatment for invasive fungal infections compared to healthy volunteers, adverse side effects included nausea, vomiting, headache, abdominal pain, and diarrhea [1]. In clinical trials, very low percentage of the following side effects were seen [60]:

(a) Hypersensitivity, hypotension, headache, dizziness, and confusion.
(b) GI symptoms, including constipation, dry mouth, diarrhea, nausea, and vomiting (3–12% each in studies).

(c) Hepatotoxicity and liver enzyme abnormalities have been reported in <2–3% of patients.

(d) Rare complications of urinary tract infections, vaginitis, and musculoskeletal pain.

(e) Cimetidine, rifabutin, and phenytoin can decrease posaconazole plasma concentrations hence their coadministration should be avoided unless necessary to do. Coadministration with ergot alkaloids is contraindicated.

Since metabolism of posaconazole is not significant through the P450 cytochrome, there are lesser expected interactions with other drugs metabolized by P450 enzymes, such as terfenadine, astemizole, cisapride, pimozide, halofantrine, or quinidine. Coadministration with these substrates can increase plasma concentrations of posaconazole, and this could lead to prolonged QTc [45]. The inhibition of P-glycoprotein by itraconazole and posaconazole may lead to increased system exposure of drugs affected by this transport system. Because interactions caused by itraconazole have been more extensively studied than those with posaconazole, drugs with known interactions to itraconazole should be used with caution in posaconazole-treated patients [60].

Posaconazole is a Category C drug and use in pregnancy should be avoided.

Resistance

Posaconazole is less active in vitro against fluconazole resistance *Candida* sp., especially *C. glabrata* and *C. krusei*. Literature on the mechanism of resistance to posaconazole are scarce.

MIC Interpretive Criteria

There are no CLSI interpretive guidelines for posaconazole.

Comments

Posaconazole has lower nephrotoxicity, hepatic toxicity, and ocular toxicity than other triazoles; since the major metabolism is performed in the liver, posaconazole has less effects on patients with renal impairment.

1.3.2.5 Ravuconazole (BMS-207147;ER-30346)

Brand Name/Formulations

There is no brand name given as yet; in Phase II trials; oral formulation only; structurally looks like itraconazole. Bristol-Myers Squibb is the manufacturer.

Primary Uses

Unknown as yet

Spectrum of Activity

Limited published data, but in vitro studies demonstrate efficacy of ravuconazole against *Candida* spp., *Aspergillus fumigatus*, *A. flavus*, *A. terreus*, and other *Aspergillus* spp., *C. neoformans*, *Zygomycetes*, *S. apiospermum*; variable to weak activity vs. *Fusarium* and *Paecilomyces* spp [1].

Mechanism of Action

Ravuconazole is an azole, and as such, it inhibits cytochrome P450-dependent enzyme lanosterol C $_{14}$ α-demethylase. This leads to disruption of membrane synthesis, depletion of ergosterol that causes an increase in toxic methylated sterol precursors in the membrane. The newer azoles, like ravuconazole also inhibit 24-methylene dihydrolanosterol demethylase to give increased activity against moulds. The action is often fungicidal vs. moulds, increasing the sensitivity of the fungus to oxygen-dependent microbicidal systems of the host.

Pharmacokinetics: (Absorption/Distribution/Excretion)

Oral bioavailability has been shown to be 47–74%; absorption rate is 4 h after oral administration, and steady state is achieved after 29 days; 98% of the drug is protein-bound; penetration into CSF has been shown to be 10% in a rabbit model; penetration in vitreous fluid in a rabbit model has been 15% of serum levels. Tissue concentrations in rabbits were high in liver, adipose tissue, bone marrow, lung, kidney, and brain [1].

Half-life is 5–7 days [1]. Few CYP450 interactions since the metabolism is only by CYP3A4 hepatic enzymes. No dose adjustments needed in patients with renal or hepatic dysfunctions.

Like other azoles, ravuconazole works with dependent kinetics; the post-antifungal effect is long. The pharmacodynamic target for efficacy is to achieve an AUC (area under the curve)/MIC of ≥25.

Adverse Reactions/Drug Interactions

There are few CYP450 interactions as compared to other azoles. Very limited data is available on drug interactions and adverse effects in humans. In a Phase I/II randomized, double-blinded study of the treatment of 148 patients with subungual onychomycosis with ravuconazole vs. a placebo, the proportion of patients with side effects was equal between the two groups. Headache and abdominal pain were most frequent side effects in 6% of patients. Three patients discontinued the drug because of dizziness, urinary incontinence, diarrhea, and malaise from anemia. Slight electrolyte abnormalities were noted in 15% of patients, and mild elevations of liver function tests were noted in 26% of study patients; 7.4% of patients had moderate to severe elevations of liver function tests [1].

Resistance

No data is yet available about any in vitro resistance.

MIC Interpretive Criteria

Not available yet

Comments

It is unknown when this may be approved.

1.3.2.6 Albaconazole

Brand Name/Formulations

No brand name given as yet

Primary Uses

Possibly for vulvovaginal candidiasis

Spectrum of Activity

There is very limited published data, but in vitro studies have demonstrated efficacy of albaconazole against *Candida* spp., *Scedosporium prolificans*, and variable activity against *Paecilomyces* spp. [1].

Mechanism of Action

Albaconazole is an azole, and as such, it inhibits cytochrome P450-dependent enzyme lanosterol C $_{14}$ α-demethylase. This leads to disruption of membrane synthesis, depletion of ergosterol that causes an increase in toxic methylated sterol precursors in the membrane. The newer azoles also inhibit 24-methylene dihydro-lanosterol demethylase to give increased activity against moulds. The action is often fungicidal vs. moulds, increasing the sensitivity of the fungus to oxygen-dependent microbicidal systems of the host.

Pharmacokinetics: (Absorption/Distribution/Excretion)

Pharmacokinetic data is only available in animals and a few healthy volunteers or patients with vulvovaginitis thus far. In dogs, the bioavailability was shown to be 100%. A CSF penetration of about 15% was seen in rabbits [1].
     Half-life is 3 days [1]. Few CYP450 interactions since the metabolism is only by CYP3A4 hepatic enzymes.

Adverse Reactions/Drug Interactions

Only preliminary data is available on the potential side effects of albaconazole. No side effects were noted in a study with healthy volunteers and patients with vul-vovaginitis, but more studies will be needed to verify this [1].

MIC Interpretive Criteria

Not available yet

Comments

It is unknown when this may be approved.

1.4   The Echinocandins

The echinocandins are one of the newest classes of antifungal agents that act by inhibiting cell wall synthesis. They are semisynthetic lipopeptides that inhibit the synthesis of 1,3-β-D-glucan, an essential cell wall component of many fungi. They specifically target the FKS1 genes in fungal organisms that encode for the compo-nents of the enzyme glucan synthase, an enzyme necessary for synthesis of 1,3-β-D-glucan, an essential component of the cell wall of susceptible fungi. Inhibition of this enzyme by the echinocandins is predominantly in *Candida* spp. and *Aspergillus* spp. There is no activity by the echinocandins against *Cryptococcus neoformans*, the Zygomycetes, the yeast forms of the endemic systemic mycoses, and other non-*Aspergillus* moulds. The β-1,3-D- glucan in these fungi is not a prominent cell wall component, if present at all, and hence no echinocandin activity is seen in vitro or in vivo. The action of the echinocandins against many susceptible fungi is considered to be fungicidal; however, against *C. parapsilosis* and *C. guilliermondii*, and *Aspergillus* spp., the echinocandins are only fungistatic [51, 53].

Echinocandins also possess immunomodulatory effects. By exposing β-glucan by the disruption of the fungal cell wall mannoproteins, additional antigens are exposed for antibody deposition and fungal recognition by the host immune system [53].

There are three echinocandins for treatment of susceptible fungal isolates in order of their FDA approval and availability in the U.S.A: caspofungin, micafungin, and anidulafungin. There are differences between them that will be described below; however, they are considered to be similar in mode of action, in vitro MIC testing and results for susceptible organisms, and in possible resistance development. Echinocandin efficacy is predicted by peak to MIC ratios, and optimal fungicidal activity is obtained when peak concentrations exceed MICs by five- to tenfold. There is poor oral absorption of the echinocandins, and thus only IV formulations are available.

There are few reported adverse side effects of the echinocandins because the glucan target is not found in mammalian cells. Resistance development has been rare to the echinocandins. A survey of >8,000 *Candida* isolates between 2001 and 2004 from >90 medical centers worldwide showed no change in MICs as compared to data collected from 1992 to 2000 vs. 3,000 *Candida* isolates vs. echinocandins [43]. A single amino acid substitution is postulated to be all that would be needed for resistance development, and some resistance has been seen in isolates of *C. albicans, C. glabrata, C. parapsilosis*, and *C. krusei*, but in patients who were on long courses of echinocandin therapy. This resistance is theoretically possible with mutations in the genes that code for the 1,3-β-D-glucan synthase, specifically FKS1 and to a lesser extent FKS2. There are also suggestions that resistance might occur via an efflux pump in the fungal cell wall and overexpression of cell wall transporter proteins [51].

1.4.1 Caspofungin

1.4.1.1 Brand Names/Formulations

Cancidas® (Merck); IV formulation

1.4.1.2 Primary Uses

FDA-approved indications for caspofungin are for the empiric treatment of presumed fungal infections in febrile neutropenic patients; candidemia; treatment of these candidal infections: intra-abdominal abscesses, peritonitis, and pleural space infections; treatment of esophageal candidiasis and for treatment of invasive aspergillosis in patients refractory to other therapies.

1.4.1.3 Mechanism of Action

The echinocandins are cyclic lipopeptides. They inhibit $1,3-\beta-D$-glucan synthase that is needed for the synthesis of cell wall glucans which act to provide structural integrity and osmotic stability to the fungus. This will prevent >90% of glucose incorporation into glucan in cell wall. The action of the echinocandin in inhibiting cell wall synthesis results in lysis of the cells [27, 51].

1.4.1.4 Spectrum of Activity

Caspofungin is active against azole susceptible and resistant strains of *Candida* sp. (*Candida parapsilosis* demonstrates highest MICs); some moulds including some species of *Aspergillus* sp.; there is activity of the echinocandins against the mycelial forms of the dimorphic fungi *Histoplasma capsulatum, Blastomyces dermatitidis, and Coccidioides immitis,* but only limited activity vs. yeast forms of these fungi; poor activity is demonstrated against Zygomycetes and *Fusarium* and *Pseudallescheria boydii*; no activity against *C. neoformans* [27, 51, 53].

1.4.1.5 Pharmacokinetics: (Absorption/Distribution/Excretion)

Caspofungin, like all of the echinocandins, is a large molecule with poor oral bio-availability. Less than 10% of the drug is absorbed, hence the use of only IV formulations. Caspofungin is highly protein-bound (up to 97%) and extensively distributed in tissues; distribution is the dominant mechanism influencing plasma clearance. Like the other echinocandins, it is easily distributed to viscera like liver and spleen, GI tract and gall bladder, and the lungs. It is less well distributed to the eyes, to the CSF, and to the brain [10, 27].

The half-life of caspofungin is 9–11 h [27]. Metabolism occurs via hydrolysis and N-acetylation in the liver; caspofungin and its inactive metabolites are excreted in feces (35%) and urine (41%), with only about 1.4% excreted unchanged in the urine. Caspofungin is a poor substrate for the cytochrome P450 enzymes [10, 27, 51]. No dosage adjustment is needed in renal or hepatic insufficiency.

Caspofungin, like other echinocandins, has a concentration-dependent pharmacokinetics. It is dosed once daily and has a long post antifungal effect against *Candida* spp. [51].

1.4.1.6 Pharmacodynamic Target

The pharmacodynamic target is a C_{max}/MIC (C = concentration of the drug) of 3 for 50% efficacy and >10 for maximal efficacy.

1.4.1.7   Adverse Reactions/Drug Interactions

The glucan target for echinocandins is not found in mammalian cells, hence decreasing the toxic side effects of these agents. Adverse side effects have been demonstrated as follows:

(a) Fever, nausea and vomiting, phlebitis, and flushing; anaphylaxis has been reported rarely [51].
(b) Laboratory abnormalities have included increased liver function tests, eosinophilia, increased proteinuria, and decreased potassium levels.
(c) Interactions have been demonstrated with rifampin; antiepileptics, including carbamazepine and phenytoin; immunosuppressants, including cyclosporine (increases caspofungin levels by 35%), tacrolimus (reduces tacrolimus levels by 20–25%), and dexamethasone; and some antiretrovirals [53].
(d) Levels of caspofungin are reduced with coadministration of dilantin, rifampin, and nevirapine [10, 53].

Caspofungin should not be given to patients who have a hypersensitivity to any component of the agent.

Concentrations of tacrolimus are decreased by ~20% when given with caspofungin; cyclosporine increases the area under the curve (AUC) of caspofungin; rifampin can decrease steady-state plasma concentration of caspofungin. There is no data on changes in presence of sirolimus, nifedipine, or voriconazole [27].

1.4.1.8   Resistance

Primary resistance in *C. neoformans* because it does not have the 1,3-β-D-glucan synthase target. Caspofungin is not active in vitro vs. *Fusarium* sp., *Rhizopus* sp., *Paecilomyces lilacinus,* nor *Scedosporium prolificans*. Mutations in the FKS1 gene that encodes glucan synthase and the GNS1 gene that encodes an enzyme involved in fatty acid elongation can theoretically result in resistance; a single amino acid substitution has the potential to lead to resistance [27, 51]. Other proposed mechanisms of resistance include the presence of an efflux pump in the cell wall and overexpression of cell wall transport proteins [51]. Surveillance studies examining the in vitro results of all three echinocandins failed to show evidence of emerging resistance over a 6-year period from 2001 to 2007. However, there have been a few reports of isolated cases of acquired resistance to echinocandins. This has not been described except in patients on very extensive courses of therapy [43].

1.4.1.9   MIC Interpretations

There are interpretive criteria available for the susceptibility of the echinocandins including caspofungin. The methods presently used for in vitro testing are those described for other antifungals, that is determination of the MIC [12]. However, there is literature supporting the use of an MEC (minimum effective concentration) in which changes in morphology are microscopically determined to suggest presence of antifungal activity [2, 16].

A nonsusceptible isolate of yeast may be indicated when the MIC to caspofungin is >2 μg/ml; however, there is no resistance breakpoint because so few isolates have demonstrated resistance.

1.4.1.10   Comments

Caspofungin is very effective in vitro against all *Candida* sp. No activity against *Cryptococcus neoformans*.

1.4.2   Micafungin

1.4.2.1   Brand Name/Formulation

Micamine™; IV formulation only

1.4.2.2   Primary Uses

For the treatment of patients with candidemia, acute disseminated candidiasis, *Candida* peritonitis, and abscesses; treatment of patients with esophageal candidiasis; prophylaxis against *Candida* infections in patients undergoing hematopoietic stem cell transplantation. It has not been FDA-approved for use in empiric therapy of the febrile neutropenic patient or for patients intolerant of or refractory to other therapies for invasive aspergillosis [51].

1.4.2.3   Mechanism of Action

Micafungin is an echinocandin; semisynthetic water-soluble lipopeptide. It is synthesized by a chemical modification of a fermentation product of *Coleophoma empetri* F-11899. Like other echinocandins, micafungin inhibits production of 1,3-β-D-glucan synthase that is needed for the synthesis of cell wall glucans which provide the structural integrity and osmotic stability to the fungus. The action of the echinocandin in inhibiting cell wall synthesis results in lysis. Micafungin exhibits fungicidal activity against *Candida* spp. and fungistatic activity against *Aspergillus* spp. [51].

1.4.2.4   Spectrum of Activity

Micafungin has good activity against *C. albicans, C. glabrata, C. krusei, C. tropicalis, and C. parapsilosis* and has been shown, including those strains that are azole resistant. There is activity against dimorphic fungi, especially the mycelial forms of these organisms. Micafungin has been shown to possess activity against *Aspergillus* spp. including *A. fumigatus* and *A. terreus.* Micafugin demonstrates activity against the mycelial forms of the dimorphic fungi *Histoplasma capsulatum, Blastomyces dermatitidis, and Coccidioides immitis,* but only limited activity vs. yeast forms of these fungi; poor activity is demonstrated against Zygomycetes and hyaline moulds including *Fusarium* and *Pseudallescheria boydii*; no activity against *C. neoformans* [27, 51].

1.4.2.5   Pharmacokinetics: (Absorption/Distribution/Excretion)

All of the echinocandins, including micafungin, have a concentration-dependent pharmacokinetics with a long post-antifungal effect. The micafungin area under the curve (AUC) is linearly related to dose over the range of 50–150 mg or 3–8 mg/kg; 85% of the steady-state concentration is reached after three daily micafungin doses [10, 27, 51].

Micafungin is >99% protein-bound, primarily to albumin and to a lesser extent, alpha-1-acid-glycoprotein. The volume of distribution is 0.39 L/kg body weight [1, 2]. Like the other echinocandins, it is easily distributed to viscera like liver and spleen, GI tract and gall bladder, and the lungs. It is less well distributed to the eyes, to the CSF, and to the brain [10]. The half-life of micafungin is ~13 h [27].

Micafungin is metabolized by nonoxidative metabolism within the liver by aryl-sulfatase and then catechol-O-methyltransferase (COMT) to form two inactive metabolites; primary elimination of micafungin and its metabolites in the stool is ~40% (although some authors suggest 71% [38]) and <15% in the urine. Micafungin does not require dosage adjustment with patients who have renal or hepatic impairment [10, 51, 53].

Micafungin exhibits linear pharmacokinetics after IV administration.

1.4.2.6   Pharmacodynamic Target

The pharmacodynamic target is a C_{max}/MIC, where C = concentration of the drug, of 3 for 50% efficacy, and >10 for maximal efficacy.

1.4.2.7   Adverse Reactions/Drug Interactions

The only contraindication is a known hypersensitivity to any component of the product. Serious hypersensitivity reactions, including anaphylaxis and anaphylactoid reactions, have been reported. Infusions should be immediately stopped and appropriate treatment administered [27].

Side effects include:

(a) There have been elevations in liver function tests, BUN, and creatinine, and rarely intravascular hemolysis and hemoglobinuria.
(b) Rash, pruritis, facial swelling, and vasodilatation have been reported.
(c) Injection site reactions including phlebitis and thrombophlebitis have been associated with peripheral IV administrations [51].
(d) In addition, other drug-related adverse effects that have been reported in patients with esophageal candidiasis include headache, leukopenia, nausea, rigors, abdominal pain, and pyrexia. The incidence of these was found in ~1.2% of all patients.
(e) No dose adjustments have to be made when coadministered with mycophenolate mofetil, cyclosporine, tacrolimus, prednisolone, fluconazole, ritonavir, or rifampin.
(f) Patients on sirolimus and nifedipine should be monitored for toxicity due to possible increased levels when coadministered with micafungin. The AUC of sirolimus was increased 21% in the presence of micafungin; the AUC and peak concentration of nifedipine were increased 18% and 42% respectively, in presence of micafungin [13].

Safety and efficacy of micafungin have not been established in pediatric patients. In pregnancy, micafungin is a Category C; it has not been studied in pregnant women and should only be used if clearly needed.

1.4.2.8 Resistance

The potential for drug resistance is suggested possibly resulting from a single amino acid substitution, but thus far little resistance has occurred in *Candida* spp. Primary resistance is present in *C. neoformans* because it does not have the 1,3-β-D-glucan synthase target. Micafungin is not active in vitro vs. *Fusarium* sp., *Rhizopus* sp., *Paecilomyces lilacinus,* nor *Scedosporium prolificans*. Mutations in the FKS1 gene that encodes glucan synthase and the GNS1 gene that encodes an enzyme involved in fatty acid elongation can theoretically result in resistance. Other proposed mechanisms of resistance include the presence of an efflux pump in the cell wall and overexpression of cell wall transport proteins [51]. Surveillance studies examining the in vitro results of all three echinocandins failed to show evidenced of emerging resistance over a 6-year period from 2001 to 2007. However, there have been a few reports of isolated cases of acquired resistance to echinocandins [43]. This has not been described except in patients on very extensive courses of therapy.

1.4.2.9 MIC Interpretation

There are interpretive criteria available for the susceptibility of the echinocandins including micafungin. The methods presently used for in vitro testing are those described for other antifungals, that is determination of the MIC [12]. However, there is literature supporting the use of an MEC (minimum effective concentration) in which changes in morphology are microscopically determined to suggest presence of antifungal activity [2, 16].

A nonsusceptible isolate of yeast may be indicated when the MIC to micafungin is >2 µg/ml; however, there is no resistance breakpoint because so few isolates have demonstrated resistance.

1.4.2.10 Comments

Micafungin was FDA-approved in the U.S.A. in the Spring of 2005 for the treatment of esophageal candidiasis and the prophylaxis if *Candida* infection in hematopoietic stem cell transplantation.

1.4.3 Anidulafungin

1.4.3.1 Brand Name/Formulations

Eraxis (Pfizer); IV formulations

1.4.3.2 Primary Uses

Candidemia and other forms of *Candida* infections including intra-abdominal abscess, peritonitis, and esophageal candidiasis.

Anidulafungin has not been studied in endocarditis, osteomyelitis, and meningitis due to *Candida* and has not been studied in sufficient numbers of neutropenic patients to determine efficacy in this group. It has not been FDA-approved for prophylaxis against *Candida* in hematopoietic stem cell transplantation or for empiric treatment in the febrile neutropenic patient nor for invasive aspergillosis in the patient that is intolerant to or refractory to other therapies [51].

1.4.3.3 Mechanism of Action

Anidulafungin is a semisynthetic lipopeptide synthesized from a fermentation product of *Aspergillus nidulans*. The echinocandins are cyclic lipopeptides. Anidulafungin is a noncompetitive inhibitor of 1,3-β-D-glucan synthase that is needed for the synthesis of cell wall glucans which act to provide structural integrity and osmotic stability to the fungus. The action of the echinocandin in inhibiting cell wall synthesis results in lysis [55].

1.4.3.4 Spectrum of Activity

Candida sp., including *C. albicans, C. tropicalis, and C. parapsilosis* and *C. lusitaniae*; anidulafungin, like the other echinocandins, has activity against fluconazole-resistant *Candida* sp., including *C. krusei* and *C. glabrata.* There is some activity of all echinocandins against the mycelial forms of the dimorphic fungi *Histoplasma capsulatum, Blastomyces dermatitidis, and Coccidioides immitis,* but only limited activity vs. yeast forms of these fungi; poor activity is demonstrated against Zygomycetes and hyaline moulds, including *Fusarium* and *Pseudallescheria boydii*; no activity against *C. neoformans* [55].

1.4.3.5 Pharmacokinetics: (Absorption/Distribution/Excretion)

Systemic exposures of anidulafungin are dose proportional; steady state was achieved in healthy volunteers on the first day after a loading dose (twice the daily maintenance dose), and the estimated plasma accumulation factor at steady state is approximately two. The clearance of anidulafungin is about 1 L/h, and anidulafungin has a terminal elimination half-life of 40–50 h [27, 51, 55]. Anidulafungin is largely bound to plasma proteins in humans (84–99%) [27, 51, 55]. The half-life of anidulafungin is ~25 h (40–50 h in some studies [51]) the longest of the three echinocandins [27, 55]. It is dosed once daily as are the other echinocandins.

The pharmacokinetics of anidulafungin following IV administration are characterized by a short distribution half-life (0.5–1 h), and a volume of distribution of 30–50 L that is similar to total body fluid volume. Anidulafungin, as with other echinocandins, achieves negligible concentrations in CSF, urine, and intravitreal fluid.

Like the other echinocandins, it exhibits linear pharmacokinetics. The echinocandins have a concentration-dependent pharmacokinetics with a long post-antifungal effect.

Anidulafungin is not metabolized in the liver, instead undergoing slow enzymatic chemical degradation in the blood to form inactive metabolites; 30% inactive drug and 10% unchanged is excreted via feces, and <1% is renally excreted [10, 27, 51, 55].

Anidulafungin is not a substrate for the CYP450 system enzymes nor does it inhibit them. No clinically relevant drug-drug interactions have been identified for anidulafungin [10, 55].

1.4.3.6 Pharmacodynamic Target

Pharmacodynamic target for most *Candida* spp. is a C_{max}/MIC ratio, where C=concentration of the drug, of 3 for 50% efficacy and >10 for maximal efficacy. For *Aspergillus* spp., the efficacy is best correlated with C_{max}/minimum effective concentration

(MEC), where MEC = lowest concentration of drug that causes the formation of abnormally branched hyphal tips.

1.4.3.7 Adverse Reactions/Drug Interactions

Anidulafungin is well tolerated, and severe adverse effects are uncommon. Anidulafungin is contraindicated in persons with known hypersensitivity to anidulafungin, any component of anidulafungin, or other echinocandins. Laboratory abnormalities in liver function tests have been seen in healthy volunteers and patients treated with anidulafungin. In some patients with serious underlying medical conditions who were receiving multiple concomitant medications along with anidulafungin, clinically significant hepatic abnormalities have occurred. Isolated cases of significant hepatic dysfunction, hepatitis, or worsening hepatic failure have been reported in patients; a causal relationship to anidulafungin has not been established. Patients who develop abnormal liver function tests during anidulafungin therapy should be monitored for evidence of worsening hepatic function and evaluated for risk/benefit of continuing anidulafungin therapy [27, 51, 55].

In vitro studies showed that anidulafungin is not metabolized by human cytochrome P450 or by isolated human hepatocytes and does not significantly inhibit the activities of clinically important human CYP isoforms (1A2, 2C9, 2D6, 3A4). No clinically relevant drug-drug interactions were observed with drugs likely to be coadministered with anidulafungin [55].

Anidulafungin is a Category C drug in pregnancy. It can cross the placental barrier in rats and was detected in fetal plasma. It is not known whether anidulafungin is excreted in human milk. Long-term carcinogenicity studies have not been conducted; it was not genotoxic in in vitro studies.

1.4.3.8 Resistance

Primary resistance in *C. neoformans* because it does not have the $1,3\text{-}\beta\text{-}\text{D-glucan}$ synthase target. Anidulafungin is not active in vitro vs. *Fusarium* sp., *Rhizopus* sp., *Paecilomyces lilacinus,* nor *Scedosporium prolificans*. Mutations in the FKS1 gene that encodes glucan synthase and the GNS1 gene that encodes an enzyme involved in fatty acid elongation can theoretically result in resistance. Other proposed mechanisms of resistance include the presence of an efflux pump in the cell wall and overexpression of cell wall transport proteins [51]. Surveillance studies examining the in vitro results of all three echinocandins failed to show evidenced of emerging resistance over a 6-year period from 2001 to 2007. However, there have been a few reports of isolated cases of acquired resistance to echinocandins [43]. This has not been described except in patients on very extensive courses of therapy.

1.4.3.9   MIC Interpretive Criteria

There are interpretive criteria available for the susceptibility of the echinocandins including caspofungin. The methods presently used for in vitro testing are those described for other antifungals, that is, determination of the MIC [6]. However, there is literature supporting the use of an MEC (minimum effective concentration) in which changes in morphology are microscopically determined to suggest presence of antifungal activity [2, 16].

1.4.3.10   MIC Interpretations

The MIC for anidulafungin is in the range of <0.03–4 μg/ml; activity against susceptible fungi is fungicidal. A nonsusceptible isolate of yeast may be indicated when the MIC to anidulafungin is >2 μg/ml; however, there is no resistance breakpoint because so few isolates have demonstrated resistance thus far.

1.4.3.11   Comments

Initial studies have demonstrated that anidulafungin may be useful for treating a range of serious fungal infections, including mucocutaneous candidiasis, invasive candidiasis, azole refractory esophagitis, and possibly invasive aspergillosis, either alone or as part of a combination regimen. More studies will be needed to verify this [55].

1.5   Miscellaneous Topical and Oral Antifungal Agents

1.5.1   Griseofulvin

1.5.1.1   Brand Name/Formulations

Fulvicin®(Schering), Grisactin® (Wyeth Pharmaceuticals); oral capsules, tablets, and oral suspension

1.5.1.2   Primary Uses

Griseofulvin is primarily used for dermatophyte (ringworm) infections of skin, hair, and nails that cannot be treated with itraconazole or terbinafine.

1.5.1.3   Spectrum of Activity

Griseofulvin has fungistatic activity against *Microsporum* spp., *Epidermophyton floccosum*, and *Trichophyton* spp. It has no activity against *Candida* spp.

1.5.1.4   Mechanism of Action

The complete mechanism of action for griseofulvin is not well understood, but is known to arrest mitosis (cell division) in metaphase and cause a disorganization of spindles (similar to the action of colchicine) and chromosome scattering in anaphase; acts on actively growing fungi only.

1.5.1.5   Pharmacokinetics: (Absorption/Distribution/Excretion)

Griseofulvin is fungistatic; following oral administration, 50% is absorbed systemically in fasting patients; 100% is bioavailable. Absorption is maximized when the ultramicronized preparation is used.

Griseofulvin is deposited in keratin precursor cells which then become impervious to fungus, hence it can be effective for treatment of onychomycosis. There is a very slow cure rate with griseofulvin; it takes weeks to months of therapy before response is observed and recurrences are not uncommon [34, 57].

1.5.1.6   Adverse Reactions/Drug Interactions

(a) Hypersensitivity reactions are the most common in the form of rashes and urticaria.
(b) Some GI symptoms and thrush.
(c) Isolated nonspecific findings including headache and fatigue.

Griseofulvin decreases concentration of some anticoagulants; barbiturates may decrease the concentration of griseofulvin; alcohol effects may be potentiated by griseofulvin; contraceptive effects may be decreased by concomitant use. It is contraindicated in patients with known hepatocellular disease [57].

Griseofulvin crosses the placenta and is contraindicated in pregnancy.

1.5.1.7   Resistance

Amount of resistance in vitro is unknown.

1.5.1.8   MIC Interpretations

There are no standardized methods of testing available; microbroth dilution has been reported in some literature; however, no interpretive criteria are available.

1.5.1.9   Comments

Griseofulvin has no affect on any other fungi than the dermatophytes. Griseofulvin remains one of the agents for tinea capitis, especially when caused by *Microsporum* spp.; however, there are more safer and rapidly active drugs now available, and for other species, griseofulvin remains as an agent if other agents fail or cannot be used [34, 57].

1.5.2   Terbinafine Hydrochloride

1.5.2.1   Brand Names/Formulations

Lamisil® (Novartis Pharmaceuticals); tablets, creams, and solution formulations available.

1.5.2.2   Primary Uses

Terbinafine is primarily used for onychomycosis of the toenail or fingernail including nail infections caused by *Trichophyton mentagrophytes, T. rubrum, Epidermophyton floccosum,* and *Scopulariopsis brevicaulis.* In vitro, low MICs have been demonstrated against *C. albicans* as well.

1.5.2.3   Spectrum of Activity

Terbinafine has an effective fungicidal effect against dermatophytes: *Microsporum* sp., *Epidermophyton floccosum,* and *Trichophyton* sp.; in vitro, fungistatic activity against *C. albicans* and *C. neoformans,* and *Scopulariopsis brevicaulis* is variable [15].

1.5.2.4   Mechanism of Action

Terbinafine is a synthetic allylamine derivative that acts to inhibit ergosterol synthesis at the level of squalene epoxidase. Squalene epoxidase inhibition results in ergosterol-depleted fungal cell membranes to exert a fungistatic effect; the toxic accumulation of intracellular squalene exerts a fungicidal effect. Increased membrane permeability prevents the normal "gatekeeper" function of the cell membrane from operating [15].

1.5.2.5   Pharmacokinetics: (Absorption/Distribution/Excretion)

Terbinafine is slightly soluble in water. Oral preparations are well absorbed (>70%), and there is a 40% bioavailability due to first-pass metabolism in the liver. Topical absorption is limited. Peak plasma concentrations of 1 µg/ml appear within 2 h after a 250-mg oral dose. The AUC increases only slightly when terbinafine is taken with meals [44].

Distribution: accumulation of the drug occurs in skin, nails, and adipose tissue. The drug is highly protein-bound (>99%). About 70% is eliminated renally; renal dose adjustment is necessary. In patients with creatinine clearance (<50 ml/min) or in patients with cirrhosis of the liver, the clearance of terbinafine is decreased by 50% compared to normal volunteers.

Half-life is ~36 h. Very slow release from skin and adipose tissue and thus a terminal half-life of 200–400 h. Terbinafine is extensively metabolized; however, none of its metabolites have significant antifungal activity [15, 35].

1.5.2.6 Adverse Reactions/Drug Interactions

Terbinafine is contraindicated in individuals with a known hypersensitivity to any components of the formulation. In the site of inhibition, squalene epoxidase is only inhibited in mammalian cells at much higher concentrations than that which is needed to inhibit the fungal enzymes; hence, terbinafine does not produce many side effects.

Side effects do, however, include [35]:

(a) GI effects, malaise, fatigue, vomiting, arthralgia, myalgia, and hair loss.
(b) Dermatologic side effects, including Stevens-Johnson syndrome and progressive toxic epidermal necrolysis have been seen.
(c) Change in ocular lens and retina and taste disturbances were reported in clinical trials.
(d) Not recommended for patients with chronic liver disease or renal disease. Rare cases of liver failure, some leading to death or liver transplant have occurred with use of terbinafine in patients with and without preexisting liver disease.
(e) Isolated cases of severe neutropenia; cases of decreases in absolute lymphocyte counts were reported in clinical trials. Cautious use in patients with known or suspected immunodeficiency; monitoring complete blood counts should be considered if patients receive the drug for greater than 6 weeks.

Terbinafine is an inhibitor of the CYP450 2D6 isoenzyme, and coadministration with drugs also metabolized in this fashion, such as tricyclic antidepressants, selective serotonin reuptake inhibitors, beta-blockers, antiarrhythmics like flecamide and propafenone, and monoamine oxidase inhibitors type B should be monitored for increased levels of these other agents. Some studies show decreases in prothrombin times when coadministered with warfarin; increased clearance of terbinafine when given with rifampin. Terbinafine increases clearance of cyclosporine by 15% and clearance of caffeine by 19%; terbinafine is decreased by 33% when given with cimetidine. No drug interactions have been seen with oral contraceptives, hormone replacement therapies, hypoglycemia, theophyllines, phenytoins, thiazide diuretics, and calcium channel blockers [35].

In vitro studies with terbinafine with human liver microsomes demonstrated no inhibition of the metabolism of tolbutamide, ethinylestradiol, and ethoxycoumarin, nor did it affect the clearance of antipyrine or digoxin in healthy human volunteers.

Category B drug in pregnancy. Treatment of onychomycosis can usually be postponed until after pregnancy is completed, and this would be recommended. Terbinafine is, however, present in breast milk or nursing mothers and hence is not recommended to be given to breast-feeding mothers.

1.5.2.7 Resistance

Resistance has been rarely seen in humans thus far; in vitro experiments have shown cross-resistance in *C. tropicalis*, for example, between azoles and terbinafine due to upregulation of efflux transporter genes [6]. Six isolates of *Trichophyton rubrum* were found resistant to terbinafine; resistance to terbinafine in these *T. rubrum* isolates appears to be due to alterations in the squalene epoxidase gene or a factor essential for its activity. Usual MICs are 0.03 µg/ml in susceptible strains of *T. rubrum*; in these resistant strains, MICs were >1.0 µg/ml [17].

1.5.2.8 MIC Interpretive Criteria

There is a CLSI standard for antimicrobial susceptibility testing of filamentous fungi that includes methods for testing terbinafine vs. dermatophytes [11].

1.5.2.9 Comments

Terbinafine is primarily a drug used to treat nail infections of hands and feet, caused by dermatophytes. There is, however, increasing literature that terbinafine may show synergy in vitro against a wider variety of moulds other than dermatophytes. In one review of use of terbinafine for onychomycosis, terbinafine was said to demonstrate greater effectiveness that itraconazole, fluconazole, and griseofulvin in dermatophyte caused onychomycosis. Their suggestion was that terbinafine become the drug of choice for these infections [15].

1.5.3 Amorolfine

1.5.3.1 Brand Name/Formulations

Loceryl™ (Roche Laboratories); topical nail lacquer

1.5.3.2 Primary Uses

Amorolfine is used for dermatophyte infections, particularly onychomycosis that do not involve the nail matrix; some in vitro antifungal activity against other moulds and yeasts involved in onychomycosis.

1.5.3.3 Mechanism of Action

Amorolfine causes an alteration of fungal cell membrane, resulting in reduction of ergosterol and accumulation of sterically nonplanar sterols. Can be fungistatic or fungicidal depending upon the organism.

1.5.3.4 Spectrum of Activity

Candida sp., *Cryptococcus* sp., dermatophytes including *Microsporum* sp., *Trichophyton* sp., and *Epidermophyton floccosum*; *Scopulariopsis*; dematiaceous mould including *Alternaria, Cladosporium, Fonsecaea*, and *Wangiella;* and *Histoplasma capsulatum, Sporothrix schenckii*, and *Coccidioides immitis*.

1.5.3.5 Pharmacokinetics: (Absorption/Distribution/Excretion)

Amorolfine penetrates and diffuses through the nail plate [37]. In a study comparing amorolfine and ciclopirox, amorolfine appeared more suitable for drug delivery from lacquer applications to human nails because it penetrated into nails via a hydrophilic pathway [37].

1.5.3.6 Adverse Reactions/Drug Interactions

Amorolfine can cause some minor local reactions, such as burning, dryness of skin, scaling, itching, erythema, and weeping of skin. There is no data for use during pregnancy, so amorolfine should be avoided in the pregnant woman. There are no known interactions with other agents.

1.5.3.7 Resistance

Mechanisms of resistance are unknown.

1.5.3.8 MIC Interpretive Criteria

No methods available for performing susceptibility tests.

1.5.3.9   Comments

None.

1.5.4   Tolnaftate

1.5.4.1   Brand Name/Formulations

Tinactin®; 1% concentrations as topical creams or liquids or powder formulations

1.5.4.2   Primary Uses

Tolnaftate is active against dermatophytes, but not candidal infections of the skin. It is less effective in areas of hyperkeratotic lesions. Tolnaftate is not effective against infections of the scalp, for example, *T. tonsurans* for tinea capitis infections.

1.5.4.3   Mechanism of Action

Tolnaftate is fungicidal; is a thiocarbamate that targets cell membrane permeability [44].

1.5.4.4   Pharmacokinetics: (Absorption/Distribution/Excretion)

Tolnaftate is a topical agent; no distribution to other tissues.

1.5.4.5   Adverse Reactions/Drug Interactions

None recognized.

1.5.4.6   Resistance

Some resistant strains have been seen; cross-resistance between terbinafine and tolnaftate has been reported in six strains of *T. rubrum* [17].

1.5.4.7   MIC Interpretive Criteria

None available.

1.5.4.8 Comments

Tolnaftate is an older drug still effective in treatment of some dermatophyte infections.

1.5.5 *Butoconazole or Butoconazole Nitrate*

1.5.5.1 Brand Name/Formulations

Gynazole-1™, Mycelex®-3 and in Canada, Femstat®

1.5.5.2 Primary Uses

Butoconazole is primarily used for the local treatment of vulvovaginal candidiasis.

1.5.5.3 Spectrum of Activity

Butoconazole nitrate is an imidazole derivative that has fungicidal activity in vitro against *Candida* spp. and has been demonstrated to be clinically effective against vaginal infections due to *Candida albicans*. In vitro activity has been demonstrated against *C. glabrata* and *Saccharomyces cerevisiae* [32].

1.5.5.4 Mechanism of Action

The exact mechanism of the antifungal action of butoconazole nitrate is unknown; however, it is presumed to function as other imidazole derivatives via inhibition of steroid synthesis. Imidazoles generally inhibit the conversion of lanosterol to ergosterol, resulting in a change in fungal cell membrane lipid composition. This structural change alters cell permeability and ultimately results in the osmotic disruption or growth inhibition of the fungal cell [18].

1.5.5.5 Pharmacokinetics: (Absorption/Distribution/Excretion)

Following vaginal administration of butoconazole nitrate vaginal cream, 1.7% (range 1.3–2.2%) of the dose was absorbed on average. Peak plasma levels (13.6–18.6 ng radioequivalents/ml of plasma) of the drug and its metabolites are attained between 12 and 24 h after vaginal administration [44].

1.5.5.6   Adverse Reactions/Drug Interactions

Hypersensitivity to any component would be a contraindication for use of butoconazole. Can cause GI abdominal pain or cramping; pelvic pain, vulvar/vaginal burning, itching, soreness, and swelling. There is no data on drug interactions.

No studies are recorded in regard to its potential carcinogenicity; studies in animals revealed no problems with infertility or mutagenicity.

Pregnancy Category C. There are, however, no adequate and well-controlled studies in pregnant women. Gynazole-1® should be used during pregnancy only if the potential benefit justifies the potential risk to the fetus. Safety and effectiveness in children have not been established.

1.5.5.7   Resistance

There is an unknown potential for resistance development.

1.5.5.8   MIC Interpretive Criteria

No standard methods for susceptibility testing.

1.5.5.9   Comments

In a comparative study of butoconazole nitrate cream vs. oral fluconazole treatment of vulvovaginal candidiasis, butoconazole resulted in a shorter time period to relief (12.9 h vs. 20.7 h for time to first relief), 75% of the butoconazole patients experienced first relief of symptoms at 24.5 h vs. 46.3 h for fluconazole patients. Fewer reported adverse events were seen in the butoconazole arm of the study as well [48].

1.5.6   Butenafine Hydrochloride

1.5.6.1   Brand Name/Formulations

Mentax (Penederm Inc.); cream formulation.

1.5.6.2   Primary Uses

Butenafine is a topical agent for dermatophyte infections, tinea pedis, tinea cruris, tinea corporis, and onychomycosis that are caused by dermatophytes and *Candida albicans*. Treatment of tinea versicolor is also included as an indication for use of butenafine.

1.5.6.3 Mechanism of Action

Butenafine is a benzylamine derivative, similar in structure to the allylamines (e.g., terbinafine), except that a butyl benzyl group replaces the allylamine group [52]. Butenafine inhibits the conversion of 2,3-oxydosqualene, a reaction catalyzed by the enzyme squalene epoxidase, thus it suppresses ergosterol biosynthesis by blocking squalene epoxidation. Like the allylamines, the benzylamine derivatives block an earlier step in ergosterol biosynthesis than do the azoles. Butenafine has anti-inflammatory as well as antifungal activity. Some authors have suggested that it may in addition alter cellular sterol composition and thereby render the cell membrane susceptible to damage as well [49].

1.5.6.4 Spectrum of Activity

Butenafine is a fungicidal antimycotic, with activity against the dermatophytes, including *Epidermophyton floccosum*, *Trichophyton mentagrophytes*, *T. rubrum*, and *T. tonsurans*. In vitro activity vs. *Candida albicans* and *Malassezia furfur* is included in the package insert and online descriptions of the cream [52].

1.5.6.5 Pharmacokinetics: (Absorption/Distribution/Excretion)

Butenafine has excellent penetration into the dermis and results in high concentrations in the skin that are maintained for a long time after dosing. The concentration of butenafine achieved in the skin of a guinea pig model at 24 h after a single application was 31.5 µg/g tissue, and this increased to 8.8 µg/g at 72 h. The minimum fungicidal concentrations (MFC) of *T mentagrophytes* and *M. canis* are 0.012 and 0.05 βg/ml, so these concentrations are well above these MFC concentrations. The total amount (or% dose) of butenafine HCl absorbed through the skin into the systemic circulation is low [49].

1.5.6.6 Adverse Reactions/Drug Interactions

Butenafine is usually well tolerated. Patients who are known to be sensitive to allylamine antifungals should use Mentax® (butenafine HCl cream) Cream, 1%, with caution, since cross-reactivity may occur. Mentax® cream, 1%, should only be used topically as directed by the physician, and contact with the eyes, nose, mouth, and other mucous membranes should be avoided [44].

Butenafine has no known mutagenicity or carcinogenicity potential nor have there been any studies that show impaired fertility when used topically as directed [49]. Safety and efficacy in pediatric patients below the age of 12 years have not been studied. Use of Mentax® cream, 1%, in pediatric patients 12–16 years of age is supported by evidence from adequate and well-controlled studies of Mentax® cream, 1%, in adults.

In controlled clinical trials, 9 (approximately 1%) of 815 patients treated with Mentax® cream, 1%, reported adverse events related to the skin. These included burning/stinging, itching, and worsening of the condition. No patient treated with Mentax® cream, 1%, discontinued treatment due to an adverse event. In the vehicle-treated patients, 2 of 718 patients discontinued because of treatment site adverse events, one of which was severe burning/stinging and itching at the site of application [49]. In uncontrolled clinical trials, the most frequently reported adverse events in patients treated with Mentax® cream, 1%, were: contact dermatitis, erythema, irritation, and itching, each occurring in less than 2% of patients [49].

1.5.6.7 Resistance

No mechanisms for resistance development are known.

1.5.6.8 MIC Interpretive Criteria

No standardized methods for susceptibility testing. An evaluation of the in vitro activity of butenafine HCl using a microdilution method for determination of susceptibilities was performed against a variety of dermatophytes and yeasts. The range of MIC for dermatophytes was 0.03–0.25 µg/ml for butenafine. There was limited activity of butenafine against *C. albicans* and no activity vs. *Malassezia furfur*. When compared to econazole and butenafine, ciclopirox demonstrated the broadest in vitro activity [28].

1.5.6.9 Comments

Butenafine 1% used topically has been reported to be efficacious in randomized clinical trials for tinea pedis, tinea corporis, and tinea cruris when used for short durations of treatment. However, its efficacy for treatment of tinea versicolor, seborrheic dermatitis, and its use as an anti-*Candida* agent has not yet been fully established [49].

1.5.7 Ciclopirox

1.5.7.1 Brand Name/Formulations

Loprox® cream, Penlac™ (Aventis Pharmaceuticals); topical nail lacquer.

1.5.7.2 Primary Uses

Ciclopirox is for treatment of dermatophyte and yeasts nail infections; for use on nails and immediately adjacent skin. Loprox® cream is indicated for the topical treatment of the following dermal infections: tinea pedis, tinea cruris, and tinea corporis due to *Trichophyton rubrum, Trichophyton mentagrophytes, Epidermophyton floccosum,* and *Microsporum canis;* candidiasis (moniliasis) due to *Candida albicans;* and tinea (pityriasis) versicolor due to *Malassezia furfur.*

1.5.7.3 Spectrum of Activity

Ciclopirox is active in vitro against *Trichophyton rubrum, T. mentagrophytes, Epidermophyton floccosum, Microsporum canis,* and *Candida albicans.* The MIC range with Sabouraud agar free of metals is 0.9–3.9 µg/ml. Ciclopirox 1% cream or lotion has been reported to have fungicidal activity against *Trichophyton mentagrophytes*; cidal activity in vitro vs. *C. albicans* has also been demonstrated [22].

1.5.7.4 Mechanism of Action

Ciclopirox, a synthetic antifungal agent, is a hydroxypyridone derivative which does not affect sterol synthesis. It is not related to the azoles or allylamines. The mechanism of action involves iron chelation, which inhibits fungal growth due to the inhibition of metal-dependent enzymes responsible for degradation of peroxides in cell. In addition, ciclopirox may modify the plasma membrane of dermatophytes and *C. albicans* [22, 44].

1.5.7.5 Pharmacokinetics: (Absorption/Distribution/Excretion)

An 8% ciclopirox nail lacquer was shown to permeate well in a bovine hoof membrane model and produced dose-dependent inhibitory effects on dermatophytes and yeast; 40–60% of the applied ciclopirox penetrated during the first 6 h of application, in infected and uninfected nails, and its concentration remained above the MIC of the nail pathogens studied experimentally [54].

Pharmacokinetic studies in men with tagged ciclopirox solution in polyethylene glycol 400 showed an average of 1.3% absorption of the dose when it was applied topically to 750 cm ^{2} on the back followed by occlusion for 6 h. Penetration studies with human cadaver skin from the back with Loprox (ciclopirox) cream with tagged ciclopirox showed the presence of 0.8–1.6% of the dose in the stratum corneum 1.5–6 h after application. The levels in the dermis were still 10–15 times above the minimum inhibitory concentrations.

Biological half-life is <2 h; excretion is via the kidney. Two days after application, only 0.01% of the dose applied could be found in the urine. Fecal excretion was negligible [21].

Autoradiographic studies with human cadaverous skin have shown that ciclopirox penetrates into the hair and through the epidermis and hair follicles into sebaceous glands and dermis; a portion of the drug does remain in stratum corneum [44].

1.5.7.6 Adverse Reactions/Drug Interactions

The olamine cream is well tolerated. A mild rash may occur rarely [22]. Concomitant use of ciclopirox topical solution with systemic antifungals for onychomycosis is not usually recommended. In the controlled clinical studies with 514 patients using Loprox® cream and in 296 patients using the vehicle cream, the incidence of adverse reactions was low. This included pruritus at the site of application in one patient and worsening of the clinical signs and symptoms in another patient using ciclopirox cream and burning in one patient and worsening of clinical signs and symptoms in another patient using the cream [22]. Safety and efficacy studies have not been performed in children under age of 10 years.

Pregnancy Category B drug. There are, however, no adequate or well-controlled studies in pregnant women. Because animal reproduction studies are not always predictive of human response, this drug should be used during pregnancy only if clearly needed [21, 44].

1.5.7.7 Resistance

There is an unknown potential for development of resistance by susceptible fungal organisms during therapy.

1.5.7.8 MIC Interpretive Criteria

No standardized methods nor breakpoints have been determined. An evaluation of the in vitro activity of ciclopirox olamine using a microdilution method for determination of susceptibilities was performed against a variety of dermatophytes and yeasts. The range of MIC for dermatophytes was 0.03–0.25 µg/ml for ciclopirox. Good activity was shown vs. yeasts with an MIC range of 0.001–0.25 µg/ml. When compared to econazole and butenafine, ciclopirox demonstrated the broadest in vitro activity [28].

1.5.7.9   Comments

Ciclopirox is only for use on nails and immediately adjacent skin; not for use for ophthalmic, intravaginal, or oral administration.

1.5.8   Naftifine

1.5.8.1   Brand Name/Formulations

Naftin® (Allergan); cream or gel for topical application.

1.5.8.2   Primary Uses

Naftifine is primarily used for skin infections: it is indicated for the topical treatment of athlete's foot (tinea pedis), tinea cruris, and tinea corporis (ringworm). Not for ophthalmic use.

1.5.8.3   Mechanism of Action

Naftifine is an allylamine that suppresses the biosynthesis of ergosterol at an earlier stage of the metabolic pathway than the azoles, independent of cytochrome P450 enzymes, by inhibiting the activity of squalene epoxidase. The resulting ergosterol deficiency is accompanied by an accumulation of squalene in the fungal cell that leads to cell death [33, 42].

1.5.8.4   Spectrum of Activity

Naftifine is a fungicidal activity demonstrated against dermatophytes including *Trichophyton rubrum, Trichophyton mentagrophytes, T. tonsurans,* and *Epidermophyton floccosum,* and *Microsporum canis, M. audouinii,* and *M. gypseum.* Fungistatic activity has been shown in vitro against *C. albicans.*

1.5.8.5   Pharmacokinetics: (Absorption/Distribution/Excretion)

Naftifine is a topical agent; no tissue distribution. Naftifine can penetrate the stratum corneum in sufficient concentrations to inhibit fungi found there. Following topical applications to skin of healthy volunteers, systemic absorption of naftifine was ~4% of the applied dose [44].

Naftifine and/or its metabolites are excreted via the urine and feces with a half-life of approximately 2–3 days [44].

1.5.8.6 Adverse Reactions/Drug Interactions

During clinical trials with Naftin® cream, 1%, the incidence of adverse reactions was as follows: burning/stinging (6%), dryness (3%), erythema (2%), itching (2%), and local irritation (2%) [44]. Long-term animal studies to evaluate the carcinogenic potential of Naftin® cream, 1%, have not been performed. In vitro and animal studies have not demonstrated any mutagenic effect or effect on fertility.

Pregnancy Category B. Reproduction studies have been performed in rats and rabbits (via oral administration) at doses 150 times or more the topical human dose and have revealed no evidence of impaired fertility or harm to the fetus due to naftifine. There are, however, no adequate and well-controlled studies in pregnant women. It is not known whether this drug is excreted in human milk. Because many drugs are excreted in human milk, caution should be exercised when Naftin® cream, 1%, is administered to a nursing woman [44].

Safety and effectiveness in pediatric patients have not been established [44].

1.5.8.7 Resistance

No resistance patterns have yet been detected for this class of drugs.

1.5.8.8 MIC Interpretive Criteria

There are no established methods for determining MIC values.

1.5.8.9 Comments

Naftifine is not for treatment of vaginal yeast infections or use in the eye.

1.5.9 Econazole

1.5.9.1 Brand Names/Formulations

Spectazole® (Econazole nitrate cream) 1%; vaginal and topical creams; 1% solution, spray, and powder [18].

1.5.9.2   Primary Uses

Econazole is for the topical application for treatment of tinea pedis, tinea cruris, and tinea corporis; also used for treatment of cutaneous candidiasis and treatment of tinea versicolor. Not for ophthalmic use.

1.5.9.3   Spectrum of Activity

Econazole is effective against dermatophytes, including *Trichophyton rubrum, T. mentagrophytes, T. tonsurans, Microsporum canis, M. gypseum, M. audouini,* and *Epidermophyton floccosum*; also effective against *Candida* spp. and *Malassezia furfur*.

1.5.9.4   Mechanism of Action

Econazole is an antifungal imidazole derivative with a structure identical to miconazole; however, it differs due to absence of one chlorine ring on one benzene ring. The specific mechanism of action is largely unknown, but like other imidazoles is felt to interact with 14 α-demethylase, a cytochrome P450 enzyme necessary to convert lanosterol to ergosterol. As ergosterol is an essential component of the fungal cell membrane, inhibition of its synthesis results in increased cellular permeability causing leakage of cellular contents. Ultrastructurally, the cell wall of *C. albicans*, for example, appears distorted after treatment with econazole, and the plasmalemma and mitochondria of *M. canis* appear altered after econazole exposure [18].

1.5.9.5   Pharmacokinetics: (Absorption/Distribution/Excretion)

Econazole is bound strongly by serum proteins and cannot be used systemically. After topical application to the skin of normal subjects, systemic absorption of econazole nitrate is extremely low. Although most of the applied drug remains on the skin surface, drug concentrations were found in the stratum corneum which, by far, exceeded the minimum inhibitory concentration for dermatophytes.

1.5.9.6   Adverse Reactions/Drug Interactions

Econazole is contraindicated in individuals who have demonstrated hypersensitivity to econazole or any of the other ingredients of the cream.

The drug should be discontinued if sensitization or excessive irritation occurs. Burning, itching, erythema, and stinging may occur rarely. In pregnancy, it is listed as a Category C; not recommended in first trimester, unless essential.

1.5.9.7   Resistance

Not described in the literature.

1.5.9.8   MIC Interpretations

An evaluation of the in vitro activity of econazole using a microdilution method for determination of susceptibilities was performed against a variety of dermatophytes and yeasts. The range of MIC for dermatophytes was <0.001–0.25 µg/ml for econazole. Lesser activity was demonstrated vs. yeasts with a broader range of MICs (0.125–>0.5 µg/ml) as compared to ciclopirox. When compared to econazole and butenafine, ciclopirox demonstrated the broadest antifungal in vitro activity overall [28].

1.5.9.9   Comment

None.

1.5.10   Sertaconazole

1.5.10.1   Brand Name/Formulations

Sertaconazole nitrate cream 2%.

1.5.10.2   Primary Uses

Sertaconazole is indicated for the topical treatment of athlete's foot (tinea pedis).

1.5.10.3   Mechanism of Action

Sertaconazole is an imidazole; inhibits cytochrome P450-dependent enzyme lanosterol C $_{14}$ α-demethylase. This leads to disruption of membrane synthesis and depletion of ergosterol that causes an increase in toxic methylated sterol precursors in the membrane. The action is often fungicidal, increasing the sensitivity of the fungus to oxygen-dependent microbicidal systems of the host.

1.5.10.4   Spectrum of Activity

Sertaconazole has fungicidal activity demonstrated against dermatophytes including *Trichophyton rubrum* and *Trichophyton mentagrophytes*.

1.5.10.5   Pharmacokinetics: (Absorption/Distribution/Excretion)

Sertaconazole is a topical agent; no tissue distribution.

1.5.10.6   Adverse Reactions/Drug Interactions

After 4 weeks of treatment of 92 patients with tinea pedis interdigitalis, only 8% experienced any adverse effects; none of the side effects were considered serious [8].

Teratogenic Effects: Pregnancy Category B: Safety and effectiveness in pediatric patients have not been established.

1.5.10.7   Resistance

No resistance patterns have yet been detected for this class of drugs.

1.5.10.8   MIC Interpretive Criteria

No established methods for determining MIC values.

1.5.10.9   Comments

This is a new agent for which data is not complete, but it does appear to be safe and effective against tinea pedis interdigitalis [8].

1.5.11   Oxiconazole

1.5.11.1   Brand Name/Formulations

Oxistat (oxiconazole nitrate) cream 1%.

1.5.11.2   Primary Uses

Oxiconazole is indicated for the topical treatment of tinea pedis, tinea cruris, and tinea corporis due to *T. rubrum, T. mentagrophytes,* or *E. floccosum;* also for treatment of tinea versicolor caused by *Malassezia furfur* [44].

1.5.11.3   Mechanism of Action

Oxiconazole is an imidazole; inhibits cytochrome P450-dependent enzyme lanosterol C$_{14}$ α-demethylase. This leads to disruption of membrane synthesis and depletion of ergosterol that causes an increase in toxic methylated sterol precursors in the membrane. The action is often fungicidal, increasing the sensitivity of the fungus to oxygen-dependent microbicidal systems of the host.

1.5.11.4   Spectrum of Activity

Oxiconazole has fungicidal activity demonstrated against dermatophytes including *Trichophyton rubrum*, *Trichophyton mentagrophytes, E. floccosum*, and *M. furfur.*

1.5.11.5   Pharmacokinetics: (Absorption/Distribution/Excretion)

Oxiconazole is a topical agent; no tissue distribution.

1.5.11.6   Adverse Reactions/Drug Interactions

Of 995 patients in initial trials with use of oxiconazole, there were about 4% adverse reactions that included pruritus, burning, contact dermatitis, folliculitis, erythema, papules, fissures, maceration, rash, stinging, and nodules [44].

Drug interactions have not been systematically evaluated.

Teratogenic Effects: Pregnancy Category B: Safety and effectiveness in pediatric patients have not been established.

1.5.11.7   Resistance

No resistance patterns have yet been detected for this class of drugs.

1.5.11.8   MIC Interpretive Criteria

There are no established methods for determining MIC values.

1.5.11.9   Comments

In one study, 60% of patients treated for tinea cruris, tinea pedis, and tinea corporis showed a clearing of their fungal infection when treated with oxiconazole lotion vs. 71% with the cream. None of the patients demonstrated side effects [25].

1.5.12   Potassium Iodide, SSKI

1.5.12.1   Brand Name/Formulations

Only an oral solution.

1.5.12.2   Primary Uses

Used for cutaneous and lymphatic sporothrichosis and in the tropics for ento-mophthoramycosis. A meta-analysis was recently done to assess the efficacy of oral SSKI for treatment of sporotrichosis, and the conclusion was that the currently available evidence is insufficient to assess the potential for oral potassium iodide in the treatment of sporotrichosis. There is no high-quality evidence for or against oral potassium iodide as a treatment for sporotrichosis. Further randomized double-blind placebo-controlled trials are needed to define the efficacy and acceptability of these interventions [59].

1.5.12.3   Mechanism of Action

The exact mechanism of fungicidal/fungistatic action is unknown. KI may work against fungi by a fungicidal mechanism, or it may enhance the body's immune and nonimmune defense mechanisms. Cell degeneration has been shown by electron microscopy to occur in *Sporothrix schenckii* yeast when placed into solutions of KI. However, KI does not increase monocyte or neutrophil killing of *S. schenckii* [50].

1.5.12.4   Pharmacokinetics: (Absorption/Distribution/Excretion)

KI is a compound made of 76% of the halogen iodine and 23% of the alkali metal potassium by weight [50]. Since potassium iodide is highly soluble in water, SSKI contains 1 g KI per milliliter (ml) of solution. This is less than 100% by weight, because SSKI is significantly more dense than pure water. Because KI is about 76.4% iodide by weight, SSKI contains about 764 mg iodide per ml.

After ingestion, KI is readily absorbed in the intestinal tract and distributes rapidly through the extracellular space. Iodine concentrates in the thyroid gland, salivary glands, gastric mucosa, choroid plexus, mammary glands, and placenta; 90% is excreted in the urine. Sweat, breast milk, and feces account for the remainder of the excretion [50].

1.5.12.5 Adverse Reactions/Drug Interactions

Side effects/adverse reactions include hypothyroidism; symptoms associated with iodine, such as a brassy taste, excessive secretions of nose, saliva, and tears; sneezing, burning, and ocular irritation; and dermal lesions, acne, loss of appetite, or upset stomach during the first several days as the body adjusts to the medication. More severe side effects which required notification of a physician are fever, weakness, unusual tiredness, swelling in the neck or throat, mouth sores, skin rash, nausea, vomiting, stomach pains, irregular heartbeat, numbness, or tingling of the hands or feet [50].

1.5.12.6 Resistance

None known.

1.5.12.7 MIC Interpretive Criteria

None available.

1.5.12.8 Comments

KI has been largely replaced with newer agents such as itraconazole for the treatment of sporotrichosis. There is a case report in the literature of successful treatment of subcutaneous sporotrichosis with a combination of terbinafine for 6 months and oral KI for 2 months; the authors concluded that terbinafine and KI should be the agents of choice for treatment of subcutaneous sporotrichosis [13].

References

1. Aperis G, Mylonakis E (2006) Newer triazole antifungal agents: pharmacology, spectrum, clinical efficacy and limitations. Expert Opin Investig Drugs 15:579–602
2. Arikan S, Lozano-Chiu M, Paetznick V, Rex JH (2001) In vitro susceptibility testing methods for caspofungin against Aspergillus and Fusarium isolates. Antimicrob Agents Chemother 45:27–30
3. Arikan S, Ostrosky-Zeichner L, Lozano-Chiu M, Paetznick V, Gordon D, Wallace T, Rex JH (2002) In vitro activity of nystatin compared with those of liposomal nystatin, amphotericin B, and fluconazole against clinical Candida isolates. J Clin Microbiol 40:1406–1412

4. Atkinson BJ, Lewis RE, Kontoyiannis DP (2008) *Candida lusitaniae* fungemia in cancer patients: risk factors for amphotericin B failure and outcome. Med Mycol 46:541–546
5. Barasch A, Griffin AV (2008) Miconazole revisited: new evidence of antifungal efficacy from laboratory and clinical trials. Future Microbiol 3:265–269
6. Barchiesi F, Calabrese D, Sanglard D, Falconi Di Francesco L, Caselli F, Giannini D, Giacometti A, Gavaudan S, Scalise G (2000) Experimental induction of fluconazole resistance in *Candida tropicalis* ATCC 750. Antimicrob Agents Chemother 44:1578–1584
7. Blum G, Perkholder S, HAAS H, Schrettl M, Wurzner R, Dierich MP, Lass-Flori C (2008) Potential basis for amphotericin B resistance in *Aspergillus terreus*. Antimicrob Agents Chemother 52:1553–1555
8. Borelli C, Korting HC, Bodeker RH, Neumeister C (2010) Safety and efficacy of sertaconazole nitrate cream 2% in the treatment of tinea pedis interdigitalis: a subgroup analysis. Cutis 85:107–111
9. Burgess MA, Bodey GP (1972) Clotrimazole (Bayb 5097): in vitro and clinical pharmacological studies. Antimicrob Agents Chemother 2:423–426
10. Cleary JD (2008) Echinocandins: pharmacokinetic and therapeutic issues. Curr Med Res Opin 25:1741–1749
11. CLSI document M38A-2 (2008a) Reference method for broth dilution antifungal susceptibility testing of filamentous fungi: approved standard, 2nd edn. Clinical and Laboratory Standards Institute, Wayne, PA
12. CLSI document M27 A3 (2008b) Reference method for broth dilution antifungal susceptibility testing of yeasts: approved standard, 3rd edn. Clinical and Laboratory Standards Institute, Wayne, PA
13. Coskun B, Saral Y, Akpolat N, Ataseven A, Cicek D (2004) Sporotrichosis successfully treated with terbinafine and potassium iodide: case report and review of the literature. Mycopathologica 158:53–56
14. Cross EW, Park S, Perlin DS (2000) Cross-resistance of clinical isolates of *Candida albicans* and *Candida glabrata* to over-the-counter azoles used in the treatment of vaginitis. Microb Drug Resist 6:155–161
15. Darkes MJ, Scott LJ, Goa KL (2003) Terbinafine: a review of its use in onychomycosis in adults. Am J Clin Dermatol 4:39–65
16. Diekema DJ, Messer SA, Hollis RJ, Jones RN, Pfaller MA (2003) Activities of caspofungin, itraconazole, posaconazole, ravuconazole, voriconazole, and amphotericin B against 448 recent clinical isolates of filamentous fungi. J Clin Microbiol 41:3623–3626
17. Favre B, Ghannoum MA, Ryder NS (2004) Biochemical characterization of terbinafine-resistant *Trichophyton rubrum* isolates. Med Mycol 42:525–529
18. Fromtling RA (1988) Overview of medically important antifungal azoles derivatives. Clin Microbiol Rev 1:187–217
19. Ghannoum MA, Herbert J, Isham N (2011) Repeated exposure of *Candida* spp. to miconazole demonstrates no development of resistance. Mycoses 54:e175–e177
20. Girmenia C (2009) New generation azole antifungals in clinical investigation. Expert Opin Investig Drugs 18:1354–1384
21. Goodman AG, Goodman LS, Rall TW, Murad F (eds) (1985) Goodman and Gilman's the pharmacological basis of therapeutics, 7th edn. Macmillan Publishing Co, New York
22. Gupta AK (2001) Ciclopirox: an overview. Int J Dermatol 40:305–310
23. Herbrecht R, Rajagopalan S, Danna R, Papadopoulos G (2010) Comparative survival and cost of antifungal therapy: posaconazole versus standard antifungals in the treatment of refractory invasive aspergillosis. Curr Med Res Opin 26:2457–2464
24. Isham N, Ghannoum MA (2010) Antifungal activity of miconazole against recent *Candida* strains. Mycoses 53:434–437
25. Jerajani HR, Amladi ST, Bongale R, Adepu V, Tendolkar UM, Sentamilselvi G, Janaki VR, Janaki C, Vidhya S, Marfatia YS, Patel K, Sharma N, Cooverj ND (2000) Evaluation of clinical efficacy and safety of once daily topical administration of 1% oxiconazole cream and lotion

in dermatophytosis: an open label, non-comparative multicenter study. Indian J Dermatol Venereol Leprol 66:188–192

26. Johnson LB, Kaufman CA (2003) Voriconazole: a new triazole antifungal agent. Clin Infect Dis 36:630–637

27. Kaufman CA, Carver PL (2008) Update on echinocandin antifungals. Semin Respir Crit Care Med 28:211–219

28. Kokjohn K, Bradley M, Griffiths B, Ghannoum M (2003) Evaluation of in vitro activity of ciclopirox olamine, butenafine HCl and econazole nitrate against dermatophytes, yeasts and bacteria. Int J Dermatol 42:11–17

29. Kontoyiannis DP, Lionakis MS, Lewis RE, Chamilos G, Healy M, Perego C, Safdar A, Kantarjian H, Champlin R, Walsh TJ, Raad II (2005) Zygomycosis in a tertiary-care cancer center in the era of *Aspergillus*-active antifungal therapy: a case-control observational study of 27 recent cases. J Infect Dis 191:1350–1360

30. Korting HC, Schollmann C (2009) The significance of itraconazole for treatment of fungal infections of skin, nails, and mucous membranes. J Dtsch Dermatol Ges 7:11–20

31. Leyden J (1998) Pharmacokinetics and pharmacology of terbinafine and itraconazole. J Am Acad Dermatol 38:S42–S47

32. Lynch ME, Sobel JD (1994) Comparative in vitro activity of antimycotic agents against pathogenic vaginal yeast isolates. J Med Vet Mycol 32:267–274

33. Maeda T, Takase M, Ishibashi A et al (1991) Synthesis and antifungal activity of butenafine hydrochloride (KP-363), a new benzylamine antifungal agent. Yakugaku Zasshi 111:126–137

34. Meadows-Oliver M (2009) Tinea capitis: diagnostic criteria and treatment options. Dermatol Nurs 21:281–286

35. McClellan KJ, Wiseman LR, Markham A (1999) Terbinafine. An update of its use in superficial mycoses. Drugs 58:179–202

36. Nagappan V, Deresinski S (2007) Posaconazole: a broad-spectrum triazole antifungal agent. Clin Infect Dis 45:1610–1617

37. Neubert RH, Gensbugel C, Jackel A, Warewig S (2006) Different physiochemical properties of antimycotic agents are relevant for penetration into and through human nails. Pharmazie 61:604–607

38. Nguyen MH, Clancy CJ, Yu VL, Yu YV, Morris AJ, Snydman DR, Sutton DA, Rinaldi MG (1998) Do in vitro susceptibility data predict the microbiologic response to amphotericin B? Results of a prospective study of patients with *Candida* fungemia. J Infect Dis 177:425–430

39. Oren I (2005) Breakthrough zygomycosis during empirical voriconazole therapy in febrile patients with neutropenia. Clin Infect Dis 40:770–771

40. Ostrosky-Zeichner L, Marr KA, Rex JH, Cohen SH (2003) Amphotericin B: time for a new "gold standard". Clin Infect Dis 37:415–425

41. Peman J, Canton E, Espinel-Ingroff A (2009) Antifungal resistance mechanisms. Expert Rev Anti Infect Ther 7:453–460

42. Petranyi G, Ryder NS, Stutz A (1984) Allylamine derivatives: a new class of synthetic antifungal agents inhibiting squalene epoxidase. Science 224:1239–1241

43. Pfaller MA, Boyken L, Hollis RJ et al (2008) In vitro susceptibility os invasive isolates of *Candida* spp. To anidulafungin, caspofungin, and micafungin: six years of global surveillance. J Clin Microbiol 46:150–156

44. Physicians Desk Reference, www.pdr.net

45. Rachwalski EJ, Wieczorkiewicz JT, Scheetz MH (2008) Posaconazole: an oral triazole with an extended spectrum of activity. Ann Pharmacother 42:1429–1438

46. Richter SS, Galask RP, Messer SA, Hollis RJ, Diekema DJ, Pfaller MA (2005) Antifungal susceptibilities of *Candida* species causing vulvovaginitis and epidemiology of recurrent cases. J Clin Microbiol 43:2155–2162

47. Sanguinetti M, Posteraro B, Fiori B, Ranno S, Torelli R, Fadda G (2005) Mechanisms of azole resistance in clinical isolates of *Candida glabrata* collected during a hospital survey of antifungal resistance. Antimicrob Agents Chemother 49:668–679

48. Seidman LS, Skokos CK (2005) An evaluation of butoconazole nitrate 2% site release vaginal cream (Gynazole-1) compared to fluconazole 150 mg tablets (Diflucan) in the time to relief of symptoms in patients with vulvovaginal candidiasis. Infect Dis Obstet Gynecol 13:197–206
49. Singal A (2008) Butenafine and superficial mycoses: current status. Expert Opin Drug Metab Toxicol 4:999–1005
50. Sterling JB, Heymann WR (2000) Potassium iodide in dermatology: a 19th century drug for the 21st century—uses, pharmacology, adverse effects, and contraindications. J Am Acad Dermatol 43:691–697
51. Sucher AJ, Chahine EB, Balcer HE (2009) Echinocandins: the newest class of antifungals. Ann Pharmacother 43:1647–1657
52. Syed TA, Maibach HI (2000) Butenafine hydrochloride: for the treatment of interdigital tinea pedis. Expert Opin Pharmacother 1:467–473
53. Thompson GR, Cadena J, Patterson TF (2009) Overview of antifungal agents. Clin Chest Med 30:203–215
54. Togni G, Mailland F (2010) Antifungal activity, experimental infections and nail permeation of an innovative ciclopirox nail lacquer based on a water-soluble biopolymer. J Drugs Dermatol 9:525–530
55. Vazquez JA (2005) Anidulafungin: a new echinocandin with a novel profile. Clin Ther 27:657–673
56. Verweij PE, Howard SJ, Melchers WJG, Denning DW (2009) Azole-resistance in *Aspergillus*: proposed nomenclature and breakpoints. Drug Resist Updat 12:141–147
57. White MH (1999) Antifungal agents. In: Armstrong DA, Cohen J (eds) Infectious diseases: fungal infections. Mosby-Wolfe, London
58. Wong-Beringer A, Jacobs RA, Guglielmo BJ (1998) Lipid formulations of amphotericin B: clinical efficacy and toxicities. Clin Infect Dis 27:603–618
59. Xue S, Gu R, Wu T, Zhang M, Wang X (2009) Oral potassium iodide for the treatment of sporotrichosis. Cochrane Database Syst Rev 4:CD006136
60. Zonios DI, Bennett JE (2008) Update on azole antifungals. Semin Respir Crit Care Med 29:198–210

Chapter 2
Antifungal Susceptibility Testing: Clinical Laboratory and Standards Institute (CLSI) Methods

Annette W. Fothergill

Abstract Antifungal susceptibility testing has become an important tool for physicians faced with making difficult treatment decisions regarding treatment of patients with fungal infections. The Clinical Laboratory and Standards Institute (CLSI) has approved methods for testing of both yeast and moulds. Testing may be accomplished via macrobroth, microbroth, or disk methods. In addition to CLSI methods, industry has provided a variety of both manual and automated systems for determining antifungal susceptibility for fungi. This, combined with an expanded list of interpretive data, has elevated antifungal susceptibility testing to a level of importance as a diagnostic test.

2.1 Introduction

Antifungal susceptibility testing (AST) has been a recognized diagnostic tool for over 20 years. Despite this, interpretation of the results and determination of how best to use these results continue to cause considerable confusion. Over the past two decades, antifungal susceptibility has undergone considerable change. AST has evolved from a nonstandardized procedure that generated results lacking clinical utility to a standardized procedure that is well controlled and that gives results physicians may use to assist with making tough clinical decisions. The Clinical and Laboratory Standards Institute (CLSI), formerly the NCCLS (National Committee on Clinical Laboratory Standards), has released four standard methods for antifungal susceptibility testing including M27-A3 [1] for macrobroth and microtiter yeast testing, M38-A2 [2] for microtiter mould testing, M44-A [3] for yeast disk diffusion testing, and M51-P [4] for mould disk diffusion testing.

A.W. Fothergill, M.A., M.B.A., MT(ASCP), CLS(NCA) (✉)
Fungus Testing Laboratory, Department of Pathology, University of Texas
Health Science Center, San Antonio, TX, USA
e-mail: fothergill@uthscsa.edu

G.S. Hall (ed.), *Interactions of Yeasts, Moulds, and Antifungal Agents:*
How to Detect Resistance, DOI 10.1007/978-1-59745-134-5_2,
© Springer Science+Business Media, LLC 2012

2.2 History

In 1985, the subcommittee on AST released its first report [5]. This document, M20-CR, Antifungal Susceptibility Testing: Committee Report, was compiled from responses to a questionnaire sent to hospitals and reference laboratories. This document indicated that AST was being conducted by approximately 20% of institutions that responded but on a rather small scale and that most of the sites conducting testing were utilizing some form of broth testing. Other methods in use included agar and disk diffusion methods. It was also noted that the method variability contributed to minimum inhibitory concentrations (MICs) that could not be reliably reproduced between institutions. Given these findings, the committee concluded that it was necessary and would be of use to the medical community to proceed toward a standardized method.

Based on an evaluation of findings, it was determined that the reference method should be a broth dilution method. Having chosen this starting point, several other parameters required investigation including inoculum preparation, test medium, incubation temperature, incubation duration, and criteria for endpoint determination. The first standard was published in 1992 [6] and was a macrobroth dilution method requiring 1-ml volumes in tubes. To avoid drug-medium interactions, RPMI-1640, a totally defined medium, was selected. Optimum incubation was determined to be 35°C for 48 or 72 h depending on species. The endpoint was defined as the lowest dilution that resulted in zero visible growth for amphotericin B or in an 80% reduction in turbidity as compared to the drug-free control tube for the azoles and 5-fluorocytosine.

Subsequent publications provided for a microtiter dilution method where parameters were the same but where endpoints were defined as zero visible growth for amphotericin B or a 50% reduction in turbidity as compared to the drug-free control well for other drugs. In addition to a more user-friendly method, the newer versions of M27 provided both QC and reference MIC ranges with break points being provided for *Candida* species. To date, interpretive guidelines have only been established for 5-fluorocytosine; some azoles including fluconazole, itraconazole, and voriconazole; and the currently available candins anidulafungin, caspofungin, and micafungin. Categories for 5-fluorocytosine include susceptible (S), intermediate (I), and resistant (R) while those for the azoles include susceptible (S), susceptible-dose-dependent (SDD), and resistant (R). The susceptible-dose-dependent category relates to yeast testing only and is not interchangeable with the intermediate category associated with bacterial and 5-fluorocytosine break points. This category is in recognition that yeast susceptibility is dependent on achieving maximum blood levels. By maintaining blood levels with higher doses of the antifungal, an isolate with an SDD endpoint may be successfully treated with a given azole [1]. The candins are categorized only as susceptible or nonsusceptible. The term nonsusceptible is reserved for this group because, to date, insufficient data exists to create the resistance category.

Procedures for mould testing were released in 2002 as document M38-A [7]. Parameters were similar to those for yeast microtiter testing with the exception of the inoculum size which was increased approximately one log. Endpoint determination differed slightly, with zero visible growth considered the endpoint for amphotericin B, itraconazole, posaconazole, and voriconazole. The endpoints for the remaining azoles and 5-fluorocytosine continued to occur at the lowest concentration with a 50% reduction in growth as compared to the drug-free control well.

Realizing that the candins could not be read in the same manner as existing drugs, a new criterion for endpoint determination was required. The minimum effective concentration (MEC) was described to assist with this group [8–10]. The MEC is a more difficult endpoint to describe and only applies to mould testing. It is a result of the aberrant growth noted when many mould species come in contact with the candins. This aberrant growth is noted in the test wells but typically continues through the highest concentrations. The MEC is the lowest concentration where aberrant growth is first noted.

The time- and labor-intensive methods for antifungal susceptibility testing were difficult for many routine laboratories to incorporate into their workflow. As a result, the committee reviewed the feasibility of adopting a disk diffusion method for yeast testing. The resulting document, M44-A [4], released in 2004, provided an alternative for categorizing yeast as susceptible, susceptible-dose-dependent, or resistant without testing for MICs. This methodology utilizes Mueller-Hinton agar which is already a staple in most routine microbiology settings. Results are available in 24–48 h, and categorical placement of isolates falls very close to those determined by the MIC provided in broth methods.

2.3   Yeast Testing

Following the recruitment of several laboratories from across the United States, a preliminary standard was introduced 7 years following the initial committee report. This standard, M27-P [6], provided guidelines and stipulated the parameters that are still in effect. The most current method for antifungal susceptibility testing of yeast fungi is outlined in CLSI document M27-A3 [1]. It is important to note that only *Candida* spp. and *Cryptococcus neoformans* have been evaluated. Despite this fact, other species are frequently tested using these parameters. Current parameters for yeast testing include RPMI-1640 as the test medium; an inoculum size of 0.5–2.5×10^3 CFU/ml prepared spectrophotometrically; incubation at 24, 48, or 72 h depending on species and/or drug; and endpoint determinations of optically clear for amphotericin B or 50% reduction in turbidity for the other drugs in the microtiter system. Endpoints are slightly different when performing testing via the macrobroth method. Endpoints for AMB remain at optically clear, but endpoints for the remaining drugs are considered at the lowest concentration that results in an 80% reduction in turbidity as compared to the drug-free control tube.

Break points for the yeast are placed into one of five categories. These categories include susceptible, intermediate, susceptible-dose-dependent, resistant, and non-susceptible. Isolates with MICs in the susceptible range indicate that the isolate is inhibited by a concentration of antifungal that is typically achieved in patients being treated by a standard dose. Currently, 5-flurocytosine is the only antifungal where the intermediate category is applicable. Isolates with MICs in this category are susceptible at a concentration that may be achieved in patients being given a standard dose but that are less likely to respond to therapy than an isolate that is considered susceptible. Susceptible-dose-dependent is a category unique to antifungal testing. This category indicates that a given drug may be effective in patients that can be treated with higher than normal doses and where maximum blood or tissue concentrations can be achieved. Resistant indicates that the isolate is not inhibited by concentrations of a drug that are typically achievable in patients. With the new class of antifungals, the candins, only two categories are being considered. These categories include susceptible and nonsusceptible. Nonsusceptible is used to categorize isolates that do not fall within the susceptible range for a drug but where resistance has not yet been defined.

Endpoints may be read either visually or spectrophotometrically. Turbidity is graded from 0 to 4 with 0 indicating optically clear and 4 indicating no reduction in turbidity compared to the turbidity of the drug-free control well. Grading of the remaining numbers include 1 for slightly hazy wells, 2 when a prominent reduction in turbidity is noted (usually approximately 50%), and 3 for a slight reduction in turbidity. This grading scale may be difficult since a true 50% reduction in turbidity as determined spectrophotometrically typically is substantially more turbid than the eye recognizes when assessing a 50% reduction.

For amphotericin B, the endpoint is the lowest concentration that inhibits visual growth or an endpoint score of 0. The endpoint for the azoles, 5-fluorocytosine, and the candins is the concentration where there is a decrease in turbidity of approximately 50% or an endpoint score of 2. When read spectrophotometrically, the endpoint is determined mathematically where a score of 0 equates with an optical density typically 5% or less of the drug-free control well. The endpoint for 2 equates with an optical density between approximately 6% and 50%.

Originally, the MIC was determined at either 48 h for *Candida* sp. or 72 h for *Cryptococcus neoformans.* The standard now permits the reading of endpoints in as few as 24 h for amphotericin B, fluconazole, and the candins. The remaining drugs should be read only at 48 or 72 h in the case of *Cryptococcus neoformans.* While the recommendation for the candins permits only the 24-h time point for determining the MIC, both amphotericin B and fluconazole may be read at either 24 or 48 h for *Candida* species.

The M27-A3 document permits some deviations from the method that may be evaluated by laboratories for use in their setting. One important deviation includes the use of media other than RPMI-1640 for testing of some drugs or species. The most widely discussed modification centers around amphotericin B. Isolates tested in RPMI-1640 result in amphotericin B MICs that are very tightly clustered around 1.0 μg/ml. This does not permit the distinction between susceptible isolates and potentially resistant ones. Antibiotic medium 3 provides a wider distribution of MIC values. Isolates with low MICs can easily be distinguished from those with much higher MICs. It is critical that clinicians determine which medium is being used when evaluating amphotericin B results. Concerns have been expressed regarding lot-to-lot variability with antibiotic medium 3. This, however, has not been observed by all testing facilities.

While this method was being developed in the United States, the European community began work on a standard method as well. The EUCAST (European Community Antifungal Susceptibility Testing) method, although similar, incorporated some revisions to the CLSI method to include the addition of a higher concentration of glucose to the RPMI-1640. This addition facilitates the rate of fungal growth allowing the MIC to be determined at 24 h as opposed to the original M27-A-mandated 48 h. Studies have shown that the two methods are equivalent despite these differences [11] and that a given set of isolates can expect the same categorical placement regardless of the method utilized.

Great interest in acquired resistance has surfaced regarding *Candida* species. Some feel that the widespread use of fluconazole has led to decreased susceptibility of *Candida* sp. to not only this agent but others within the azole class revealing cross-resistance. Although it is possible to find azole resistance in any given collection or clinical setting, such resistance is not as widespread as some may fear (Table 2.1). Species resistance can be assessed by the MIC_{50} and MIC_{90}. These two values represent the MIC at which 50% or 90% of the isolates tested fall at or below. It is not to be confused with the mean MIC nor the median MIC but rather is a reflection of the MICs obtained for a test set of isolates.

Despite the presence of resistance in the clinical setting, the MIC_{50} and MIC_{90} for most species fall within what is considered a susceptible range. Notable exceptions include *Candida glabrata* against the azoles, *Candida krusei* against fluconazole, and *Cryptococcus neoformans* against caspofungin. It is critical to note that caspofungin is not recommended for *C. neoformans* nor is fluconazole recommended for *Candida krusei*. Neither of these isolates should be tested against the respective drugs to which they are intrinsically resistant.

Early reports have described *Candida lusitaniae* resistance to AMB and have shown that this species possesses the possibility of developing resistance while the patient is on treatment. The first report involved a patient whose initial isolate was susceptible but whose subsequent isolates had developed AMB resistance [12]. Later reports have shown AMB resistance may exist even prior to exposure to AMB [13]. The expected rate of resistance for *C. lusitaniae* is 8–10% of any given stock collection.

When discussing utility of susceptibility testing and its correlation to patient outcome, it is best to reference the document by Rex and Pfaller [14]. Some assumptions may be made regarding the MIC and patient outcome. Rex and Pfaller propose the "90–60 Rule." This rule states that infections caused by isolates that have MICs considered susceptible respond favorably to appropriate therapy approximately 90% of the time, whereas infections caused by isolates with MICs considered resistant respond favorably approximately 60% of the time.

Table 2.1 Susceptibility patterns for clinical yeast isolates collected from 2000 to 2009

		AMB	CAS	5FC	FLU	ITRA	VORI
C. albicans	N tested	1,878	2,089	465	3,291	1,173	1,251
	MIC range	0.06 to 2.0	≤0.03 to ≥16	≤0.125 to ≥64	≤0.125 to 64	≤0.015 to ≥8.0	≤0.015 to ≥8
	MIC_{50}	0.25	0.06	0.25	0.25	≤0.015	≤0.015
	MIC_{90}	0.25	0.125	2.0	1.0	0.5	1.0
C. glabrata	N tested	1,338	1,866	348	2,326	961	1,169
	MIC range	0.125 to ≥16	0.06 to ≥16	≤0.125 to 16	≤0.125 to ≥64	≤0.015 to ≥8.0	≤0.015 to ≥8
	MIC_{50}	0.25	0.125	≤0.125	8.0	1.0	0.5
	MIC_{90}	0.5	0.25	≤0.125	≥64	8.0	4.0
C. krusei	N tested	219	285	47	213	114	196
	MIC range	0.125 to 2.0	0.06 to 8.0	2.0 to 16	8.0 to >64	0.08 to 8.0	0.125 to 8.0
	MIC_{50}	0.25	0.25	8.0	32	0.25	0.5
	MIC_{90}	0.5	0.25	8.0	>64	0.5	1.0
C. neoformans	N tested	313	72	186	444	166	223
	MIC range	0.06 to 0.5	4 to >16	≤0.125 to >64	≤0.125 to 64	≤0.015 to 0.5	≤0.015 to 0.5
	MIC_{50}	0.25	16	8.0	2.0	0.06	0.06
	MIC_{90}	0.5	>16	8.0	8	0.25	0.25

MICs reported in µg/ml

AMB amphotericin B, *CAS* caspofungin, *5FC* 5-flurocytosine, *FLU* fluconazole, *ITRA* itraconazole, *VORI* voriconazole, *N tested* number tested, *MIC* minimum inhibitory concentration

Realizing that M27-A3 is very labor intensive and not easily incorporated into busy clinical settings, the CLSI introduced M44-A. This method is a disk diffusion method that is similar to the routine Kirby-Bauer method utilized globally for bacterial susceptibility testing. To date, only fluconazole and voriconazole have been standardized, but the committee has evaluated other antifungals against both yeast and moulds. This method utilizes the same Mueller-Hinton agar that is required for bacterial testing but stipulates the addition of methylene blue-glucose to assist with yeast growth and to enhance visualization of the zone diameters. While it is beneficial for yeast testing, it does not hold true when testing moulds and M51-P methods eliminate the addition of methylene blue preparations.

Methylene blue-glucose solution is added to the surface of the Mueller-Hinton agar and permitted to air dry prior to adding the yeast inoculum. Laboratories are likely to find that M44-A fits into their workflow more easily than M27-A3 and appreciate the added benefit of being much less costly. Much work has been done to provide QC limits to ensure this method has the same validity as the original M27-A3 [15].

Since approved methods have been developed, commercial products have been introduced to assist laboratories with AST. Systems that have been evaluated include the YeastOne system by Trek Diagnostics and the Etest by AB Biodisk. These methods are easy to incorporate into the routine laboratory and give equivalent results to M27-A3 [16–18]. In addition, automated methods are under development with the Vitek by bioMérieux having FDA approval for fluconazole. Prior to launching an AST program, institutions should consider the volume of testing they can expect. The method is inherently variable, and reproducibility can be a problem. Another problem is the availability of an individual to discuss interpretation of the testing with clinicians.

Clearly, antifungal susceptibility testing of yeast fungi has become routine in many settings, and physicians are relying on MIC data to assist with difficult clinical decisions. The release of testing systems by industry such as the Etest (AB Biodisk) and YeastOne panels (Trek Diagnostics, Sensititre) has made this testing a reality in routine microbiology laboratories. While many clinicians will order susceptibility testing, there continues to be much confusion regarding the use of the results. Sufficient data has been generated to suggest susceptibility trends for specific isolates against specific agents, but direct patient outcome-MIC correlation data is minimal. Despite the lack of correlation data, antifungal susceptibility testing continues to provide useful information to assist with patient care.

2.4 Mould Testing

A new mould method, CLSI M38-A2, was released in 2002. This method is nearly identical to M27-A3 with the exception of the inoculum size. The inoculum size is determined spectrophotometrically but to a higher desired final concentration of $0.4–5 \times 10^4$ CFU/ml. The guideline provides target percent transmission (%T) readings based on conidial size that are listed by species. Isolates from the genera of *Aspergillus* spp., *Paecilomyces* spp., and *Sporothrix* spp. are measured at 80–82%T while species with larger conidia such as *Fusarium* spp., *Rhizopus* spp., and *Scedosporium* spp. are standardized to 68–70%T. Efforts are under way to determine the correct%T for most of the clinically significant fungi, but the list is not yet complete. When fungi not discussed in the M38-A2 are tested, laboratories must determine the correct %T through trial and error to achieve the desired final concentration.

From several years of the use of M27 documents, it was recognized that the scientific community preferred the microtiter method to a macrobroth one. As a result, the macrobroth method is not discussed in M38-A2. This poses a problem when testing the endemic fungi such as *Histoplasma capsulatum, Blastomyces dermatitidis,* or *Coccidioides immitis.* When necessary, mould testing may be conducted by the macrobroth method as early studies have shown that the two methods are equivalent. Other fungi that may benefit from testing by the macrobroth method are those fungi that grow very slowly. It is difficult to hold microtiter tests longer than 72 h due to dehydration. Many of the less frequently encountered fungi may require as long as 120–144 h before growth is detected in the drug-free growth control well. For this reason, isolates that are known to be slow growers should be tested via the macrobroth method.

Endpoint determination is also much more difficult with moulds than with the yeast fungi. While a reduction in turbidity is typically easy to visualize with yeast fungi, it is not so easily visualized when moulds are tested. Due to the unique growth patterns of the mould fungi, one looks for a decrease in volume of growth rather than a reduction in turbidity as for the yeasts. *Aspergillus* spp., for example, growth is seen as a cottony clump in the broth. To determine an endpoint, the reader must assess the amount of growth for each concentration and call the endpoint at that concentration that has at least 50% smaller volume of growth for antifungals not read at 100% inhibition. Many individuals are not comfortable with this subjective endpoint determination and prefer to refer mould testing to reference centers.

Reading the MIC endpoint for moulds differs from the criteria established for the yeast fungi. Amphotericin B, itraconazole, posaconazole, and voriconazole endpoints are all determined at the lowest concentration that prevents discernable growth or in other words, the first clear well. Fluconazole and 5-fluorocytosine are determined at the lowest concentration that correlates with a 50% reduction in growth as the MIC. The candins do not provide a MIC but rather a MEC, or minimum effective concentration. The candins attack the growing tips of the hyphae resulting in aberrant, stubby growth of the hyphae. This aberrant growth is easily visualized as the hyphae cluster within the well in clumps. The MEC is the lowest concentration where the growth within the well is visually clumped. Microscopic examination will display obviously distorted hyphae.

Work has not been completed that permits categorizing moulds as susceptible or not. General guidelines have been established to assist with analyzing mould data. Based on large amounts of data [19], isolates are considered susceptible to amphotericin B, itraconazole, posaconazole, voriconazole, and caspofungin when the

MIC/MEC is ≤1.0 µg/ml, intermediate with MIC/MEC is 2.0 µg/ml, and resistant when the MIC/MEC is ≥4.0 µg/ml. It is likely that the other candins would fit into these ranges as well.

2.5 Conclusion

Antifungal susceptibility testing has indeed come of age. Physicians have discovered its utility and accepted its limitations when seeking assistance with tough clinical cases. While moulds are typically not tested in routine settings, yeast fungi are more frequently incorporated into clinical laboratories. As more drugs reach the market, the CLSI will be challenged to expand existing documents, especially when new classes of drugs are introduced. In the interim, the CLSI continues to monitor medical mycology to ensure appropriate methods are available for clinical testing.

References

1. CLSI (2008a) Reference method for broth dilution antifungal susceptibility testing of yeasts; approved standard-third edition; CLSI document M27-A3. Clinical and Laboratory Standards Institute, Wayne
2. CLSI (2008b) Reference method for broth dilution antifungal susceptibility testing of filamentous fungi; approved standard CLSI document M38-A2. Clinical and Laboratory Standards Institute, Wayne
3. CLSI (2009) Method for antifungal disk diffusion susceptibility testing of filamentous fungi; proposed guideline. CLSI document M51-P. Clinical and Laboratory Standards Institute, Wayne
4. NCCLS (2004) Reference method for antifungal disk diffusion susceptibility testing of yeasts; approved guideline. NCCLS document M44-A. National Committee for Clinical Laboratory Standards, Wayne
5. NCCLS (1985) Antifungal susceptibility testing; committee report. NCCLS document M20-CR. NCCLS, Villanova
6. NCCLS (1992) Reference method for broth dilution antifungal susceptibility testing of yeasts; proposed standard NCCLS document M27-P. National Committee for Clinical Laboratory Standards, Wayne
7. NCCLS (2002) Reference method for broth dilution antifungal susceptibility testing of conidial-forming filamentous fungi. Approved standard NCCLS M38-A. National Committee for Clinical Laboratory Standards, Wayne
8. Odds FC, Motyl M, Andrade R et al (2004) Interlaboratory comparison of results of susceptibility testing with caspofungin against *Candida* and *Aspergillus* species. J Clin Microbiol 42:3475–3482
9. Espinel-Ingroff A (2003) Evaluation of broth microdilution testing parameters and agar diffusion Etest procedure for testing susceptibilities of *Aspergillus* spp. to caspofungin acetate (MK-0991). J Clin Microbiol 41:403–409
10. Espinel-Ingroff A, Fothergill A, Ghannoum MA, Manavathu E, Ostrosky-Zeichner L, Pfaller M, Rinaldi M, Schell W, Walsh T (2007) Quality control and reference guidelines for CLSI broth micro-dilution method (M38-A document) for susceptibility testing of anidulafungin against moulds. J Clin Microbiol 45(7):2180–2182

11. Espinel-Ingroff A, Barchiesi F, Cuenca-Estrella M, Pfaller MA, Rinaldi M, Rodriguez-Tudela JL et al (2005) International and multicenter comparison of EUCAST and CLSI M27-A2 broth microdilution methods for testing susceptibilities of *Candida* spp. to fluconazole, itraconazole, posaconazole, and voriconazole. J Clin Microbiol 43(8):3884–3889

12. Pappagianis D, Collins MS, Hector R, Remington J (1979) Development of resistance to amphotericin B in *Candida lusitaniae* infecting a human. Antimicrob Agents Chemother 16:123–126

13. Merz WG (1984) Candida lusitaniae: frequency of recovery, colonization, infection, and amphotericin B resistance. J Clin Microbiol 20:1194–1195

14. Rex JH, Pfaller MA (2002) Has antifungal susceptibility testing come of age? Clin Infect Dis 35(8):982–989

15. Barry A, Bille J, Brown S, Ellis D, Meis J, Pfaller M et al (2003) Quality control limits from fluconazole disk susceptibility tests on Mueller-Hinton agar with glucose and methylene blue. J Clin Microbiol 41(7):3410–3412

16. Espinel-Ingroff A, Pfaller M, Messer SA, Knapp CC, Holliday N, Killian SB (2004) Multicenter comparison of the Sensititre YeastOne colorimetric antifungal panel with the NCCLS M27-A2 reference methods for testing new antifungal agents against clinical isolates of *Candida* spp. J Clin Microbiol 42(2):718–721

17. Maxwell MJ, Messer SA, Hollis RJ, Boyken L, Tendolkar S, Diekema DJ et al (2003) Evaluation of Etest method for determining fluconazole and voriconazole MICs for 279 clinical isolates of *Candida* species infrequently isolated from blood. J Clin Microbiol 41(3):1087–1090

18. Maxwell MJ, Messer SA, Hollis RJ, Diekema DJ, Pfaller MA (2003) Evaluation of Etest method for determining voriconazole and amphotericin B MICs for 162 clinical isolates of *Cryptococcus neoformans*. J Clin Microbiol 41(1):97–99

19. Espinel-Ingroff A, Arthington-Skaggs B, Iqbal N, Ellis D, Pfaller MA, Messer S, Rinaldi M, Fothergill A, Gibbs D, Wang A (2007) Multicenter evaluation of a new disk agar diffusion method for susceptibility testing of filamentous fungi with voriconazole, posaconazole, itraconazole, amphotericin B, and caspofungin. J Clin Microbiol 45(6):1811–1820

Chapter 3
Antifungal Susceptibility Testing Methods: Non-CLSI Methods for Yeast and Moulds

Audrey Wanger

Abstract There are defined methods for performing susceptibility testing that have been published by the Clinical and Laboratory Standards Institute (CLSI) for yeast and moulds. There are commercially available products for performance of the recommended broth microdilution methods that are contained in those documents. These are described in Chapter 2 of this text. In addition, there are other available methods that are described in this chapter that are culture-based and non-culture-based. These include the use of disk diffusion, Etest, chromogenic media, and molecular methods.

3.1 Background

In 1997, the National Committee of Clinical Laboratory Standards (NCCLS), now the Clinical Laboratory Standards Institute (CLSI), developed a standardized method for susceptibility testing of yeasts and later moulds [24]. Although no MIC interpretive breakpoints exist for moulds with any of the currently available antifungal agents, reproducible results have been documented in the literature with the CLSI as well as alternative methods. A major limitation of the CLSI reference method is the need to prepare the MIC broth microdilution trays, which is a very labor intensive process and requires significant resources for quality control of the reagents and procedures. Problems also exist in manual reading of the trailing endpoints seen for several drug-organism combinations in the reference broth microdilution method. The ability of the broth microdilution method to accurately detect resistance to amphotericin remains unresolved. The CLSI disk diffusion method [22] was developed as an alternative qualitative method of antifungal

A. Wanger, Ph.D. (✉)
Department of Pathology, University of Texas Medical School, Houston, TX, USA
e-mail: audrey.wanger@uth.tmc.edu

G.S. Hall (ed.), *Interactions of Yeasts, Moulds, and Antifungal Agents:*
How to Detect Resistance, DOI 10.1007/978-1-59745-134-5_3,
© Springer Science+Business Media, LLC 2012

susceptibility testing for laboratories without the resources to perform broth microdilution testing. The disk method requires the laboratory to purchase methylene blue dye and supplement their ready-made Mueller Hinton agar plates with the dye and glucose. Disks are only commercially available for fluconazole and voriconazole.

The Antifungal Susceptibility Testing Subcommittee of the European Committee on Antibiotic Susceptibility Testing (EUCAST) also has a reference method for antifungal susceptibility testing that is a modification of the CLSI method. Differences include an inoculum that is 100× higher than CLSI, and the supplementation of the RPMI 1640 broth with 2% glucose (10× higher than CLSI), and an incubation period of 24 h versus 48 h for the CLSI procedure. In a comparative study using 100 bloodstream isolates of *Candida* species, Cuenca-Estrella showed good agreement between the two reference methods [9].

For practical reasons, the CLSI reference method recommends visual reading of the broth microdilution trays. However, spectrophotometric reading of broth microdilution trays can help eliminate variability and allows for reproducible reading, especially for azoles that can show significant trailing of growth at the endpoint, and can reduce subjectivity in the selection of MIC results. The microtiter plates are agitated so that a uniform suspension is obtained, and the turbidity can be read in a spectrophotometer. Azoles and flucytosine (FC) MIC endpoints are read at 50% inhibition and amphotericin is read at 90% inhibition as compared to controls for spectrophotometric reading [28]. Good agreement was seen in comparing visual and spectrophotometric readings of 100 clinical isolates.

3.2   Broth Microdilution Methods

A commercially available broth microdilution product for antifungal susceptibility testing Sensititre YeastOne Colorimetric system is currently available (Trek Diagnostic Systems, Cleveland, OH). The clinical panel available in the United States contains fluconazole, itraconazole, voriconazole, 5-flucytosine and is cleared for testing of *Candida* species only. Addition of lyophilized dilutions of Alamar blue oxidation-reduction colorimetric indicator to the test system is intended to aid in reading trailing endpoints [12]. The MIC result is read as the first well that shows a change in color from pink to purple (indicating inhibition of growth). A research-based panel is available which contains amphotericin B (AP) fluconazole, itraconazole, voriconazole, posaconazole as well as caspofungin, anidulafungin and micafungin. Custom dried panels without Alamar blue are also available on request from Trek Diagnostic Systems. Itraconazole MIC results have been reported to be higher with this method as compared to the CLSI method, and agreement is low for the azoles for certain *Candida* species, particularly *C. glabrata* and *C. tropicalis* [1].

Overall agreement in a study of 728 isolates of *Candida* and 78 *Cryptococcus* species was 98% as compared to the CLSI method for five drugs and Candida species. Agreement for *Cryptococcus* was good (94–100%) for azoles and FC, but 74% for AP with the YeastOne panels giving lower MICs than the reference broth method [26].

BioMérieux recently received FDA clearance for an antifungal susceptibility testing card containing fluconazole for use on their automated Vitek 2 instrument. Results are reported to be available after a minimum of 13 h incubation in the instrument. Excellent essential agreement was seen between and CLSI reference method, read after 24 h of incubation, with an overall categorical agreement of 90%. [49]. In a large study with Vitek 2 for 426 clinical isolates of *Candida* sp., the categorical agreement with broth microdilution was 88% at 24 h and 97% at 48 h [27]. A 64-well investigational card with amphotericin, flucytosine, fluconazole, and voriconazole is also available for the Vitek 2.

Other broth-based commercial methods that are available outside of the United States include the Fungitest (Sanofi Diagnostics, Pasteur, Paris, France), a broth microdilution system with breakpoint dilutions AP (2, 8), FC (2, 32), miconazole (0.5, 8), ketoconazole (0.5, 4), itraconazole (0.5, 4), and FL (8, 64) in modified RPMI broth with a color-indicating dye. The final inoculum concentration is a 1×10^3 CFU/ml. Interlaboratory agreement varied between 56% and 100%, with results for the azoles demonstrating the poorest agreement [47]. Results which showed good interlaboratory agreement were compared to the CLSI method and agreement was 56–100%, again with poor results for all the azoles. Davey also showed poor agreement for Fungitest with the azoles, particularly for *C. glabrata* with agreement as low as 38–56% [10]. Sixteen percent of FL-resistant isolates were reported as falsely susceptible by this method. The authors stated that the method required further development. Other methods which have undergone limited investigations and in general demonstrate poor agreement with the reference method include the Candifast (International Microbio/Stago, Milan, Italy), with eight wells for biochemical identification and a single well each for AP (4 µg/ml), FC (35 µg/ml), econazole, ketoconazole, miconazole, and FL at 16 µg/ml. Integral System Yeast (Liofilchem Diagnostics, L'Aquila, Italy) available only outside of the United States comprises wells for identification and the following drugs: nystatin (200 units/ml), AP (200 µg/ml), FC 20 µg/ml, econazole, ketoconazole, and FL at 100 µg/ml. A study in Italy of 800 *Candida* isolates compared the CLSI reference method to Etest, disk diffusion, Sensititre YeastOne, Fungitest, Candifast, and Integral System Yeast for testing of fluconazole. Overall agreement was 78–82% except for Candifast and Integral System which showed poor agreement (22–37%) due to a lack of standardization in inoculum preparation and medium. The authors recommended that these methods should be avoided for clinical and epidemiological studies [20]. ATB Fungus (API-bioMerieux, Marcy l'Etoile, France) uses Yeast Nitrogen Base (YNB) and includes FC (0.25–128 µg/ml), AP (1–8 µg/ml), and miconazole, ketoconazole, and econazole (1–8 µg/ml). This method recommends an inoculum suspension with a turbidity equivalent to a McFarland 2.0 (100 µl added to each cupule), incubation at 30°C for 48 h, and the endpoint criteria of turbidity present or absent. Correlation of ATB Fungus with an agar dilution procedure using YNB agar showed poor agreement [33]. Other methods for which no comparative data could be found include Mycostandard (Institute Pasteur, France), Mycototal (Behring Diagnostic, Rueil-Malmaison, France), and ASTY (Kyoto Tokyo).

Bioscreen microdilution, a semiautomated computer-controlled instrumentation, is available from Labsystems, Helsinki, Finland. This method is based on turbidimetric reading of growth in a microdilution format and generation of growth curves. These growth curves are used to provide an extrapolated MIC result and also allow rates of growth inhibition to be assessed. Excellent agreement was seen with the CLSI method [44].

More rapid generation of results can also be achieved by the addition of chemical compounds to the broth media. Consumption of these compounds during growth of the fungus can be detected more rapidly than measuring growth of the fungus. XTT (tetrazolium salt), an electron transport agent, has been used in these test systems. A concentration-dependent increased rate of XTT conversion to a reduced metabolite allows for early detection of fungal growth (as early as 6 h for zygomycetes). Good agreement was seen between the XTT rapid assay method and the CLSI reference method [2].

Addition of carboxyfluorescein diacetate (CFDA), a novel fluorescent dye, to microdilution trays after incubation was also found to aid rapid detection of antifungal resistance and allow for more objective reading of MIC endpoints. Microdilution trays read visually at 24–48 h were compared to reading using a fluorescent reader. MIC values of FL for 68 strains of *Candida* sp. were compared to the CLSI broth microdilution method and demonstrated a correlation of 97.6% at 24 h [17]. The advantage of adding the dye to the microdilution tray after incubation of the yeast with the antifungal agent is to ensure that no interference is caused by the dye itself. The dye diffuses across the cell membrane and is hydrolyzed to the fluorescent component; the dye then leaks out of the damaged membranes and is measured.

Another dye that has been added to broth microdilution trays is chloromethylfluorescein diacetate which is cleaved into a fluorescent product in the growing cell. The dye was added to microdilution plates after 16 h of incubation of *Aspergillus* species, and fluorescence was read and compared with that of drug-free controls (100% reduction in fluorescence for amphotericin and 90% reduction for the azoles). Preliminary data using two-well characterized strains indicate good correlation with the CLSI reference method [4].

3.3   Agar Methods

The broth microdilution method for testing of caspofungin has met various in vitro limitations. Problems have been observed in performance of broth microdilution MICs for caspofungin. Specifically for some isolates, growth may be observed in all of the wells (trailing), and this does not appear to correlate with treatment outcome of patients [15]. An agar dilution assay for testing of the echinocandins with *Aspergillus* species has also been developed. These as well as other investigators have proposed the use of the MEC or minimum effective concentration or the lowest concentration at which the fungi display microscopic morphological changes (correlated with secretion of galactomannan) as a measure of the antifungal activity instead of the standard MIC result [14]. Agar dilution MICs were within one dilution of MECs for 85% of isolates tested.

3.3.1 Disk Diffusion

In addition to the reference broth microbroth method, the CLSI has also published a standard for antifungal susceptibility testing using the disk diffusion method [22] and with interpretive criteria for fluconazole, voriconazole and caspofungin. Outside of the United States, a modification of the disk diffusion method that comprises compressed 9-mm tablet (Neo-Sensitabs) (Rosco Diagnostica, Taastrup, Denmark), with the antifungal agent has been in use for several decades. Neo-Sensitabs tablets are available for a variety of antifungal agents. The tablets are applied to an inoculated agar plate made with Shadomy medium, a modification of Yeast Nitrogen base, as recommended by the manufacturer and then incubated for 24–48 h. Inhibition zone diameters are read and interpreted according to criteria provided by the manufacturer. Studies of the tablet method for both yeast and mould species in comparison to CLSI broth microdilution and disk methods have been performed [11], and categorical agreements of 70–80% were achieved for posaconazole and 50–70% for amphotericin. A study in Belgium using fluconazole Neo-Sensitabs found too many major errors using interpretive criteria provided by the manufacturer, compared to the CLSI reference method to justify its use [45].

3.3.2 Direct Testing on Chromagar

Direct inoculation of blood from positive blood culture bottles for yeast onto Chromagar for the identification of *Candida* and placement of an FL disk (25 μg) on the agar has also been evaluated, to achieve both yeast identification and susceptibility results simultaneously. All isolates tested (95), except one *C. krusei*, were FL S and agreed with the CLSI method. There were two very major errors for results were read at 24 h compared to broth microdilution, a *C. glabrata* and *C. parapsilosis*. Overall, there was also a tendency for the standardized disk diffusion method to undercall resistance when *C. glabrata* isolates were read after 24 h of incubation [43].

3.3.3 Etest for MIC Determinations

Etest is an agar-based MIC method for antifungal susceptibility testing that is approved by the FDA for clinical use in the United States. This innovative gradient technique introduced in 1988 [8] illustrated an application for MIC testing of fungi. Etest comprises a preformed and predefined gradient of antibiotic or antifungal concentrations, immobilized in a dry format onto the surface of a plastic strip. The concentration gradient is calibrated across an MIC range corresponding to 15 two-fold dilutions. When applied to the surface of an inoculated agar plate, the antibiotic or antifungal agent on the Etest strip is instantaneously transferred to the agar in the form of a stable and continuous gradient directly beneath and in the immediate vicinity of the strip. The stability of the gradient is maintained for up to 18–20 h which covers the critical times of a wide range of pathogens, from rapid-growing aerobic bacteria to slow-growing fastidious organisms including fungi. The stable gradient also provides inoculum tolerance where 100-fold variation in CFU/ml has minimal effect on the MIC result itself for homogeneously susceptible strains. Thus, inoculum variability seen in routine susceptibility testing will have minimal effects on Etest MIC results unlike other susceptibility testing methods. More importantly, the stable gradient also allows the use of a macromethod with heavier inoculum to optimize the detection of low-level resistance, heteroresistance, and resistant subpopulations.

When using Etest, exact MIC results for a wide variety of antifungal agents and for most yeast can be read after 24 h of incubation, and for moulds, after 24 h to 4–5 days. Etest is FDA-cleared for testing of fluconazole, voriconazole, flucytosine, itraconazole, and micafungin. Investigational strips are available for ketoconazole, amphotericin B, and newer agents such as posaconazole, caspofungin, and anidula-fungin. Etest is thus far the only method that accurately detects amphotericin resistance [46]. Media recommended by the manufacturer for use with Etest is RPMI 1640 (same medium as for the CLSI broth method) but supplemented with 2% glucose and buffered with MOPS. Studies involving different media with Etest, e.g., modified casitone and antibiotic medium 3 agar, have shown that RPMI gave the best performance in comparison to the CLSI method [23].

Trailing of growth that is observed in the CLSI broth microdilution method as well as other methods is also seen with Etest in the form of microcolonies growing within an otherwise discernable ellipse. In a study by Pfaller et al., agreement between Etest and CLSI method was best achieved by ignoring the growth within the ellipse when reading the MICs for azole antifungal agents [31].

Etest was found to be a reproducible method in a study by Barry et al. using 50 challenge strains tested in triplicate in three different laboratories with >90% agreement was seen for fluconazole MIC values within ±1 dilution of the mode [7].

3.3.3.1 Comparison of Etest and CLSI Methods

Comparisons between Etest and CLSI methods have achieved >90% agreement for testing of *Candida* species against fluconazole, voriconazole, posaconazole [40], amphotericin, flucytosine, and caspofungin [30] and excellent agreement between Etest and broth microdilution in a study with 162 strains of *Cryptococcus neoformans* for voriconazole and amphotericin (94% and 99%, respectively) [18].

3.3.3.2 Comparison of Etest and Other Microdilution Broth Methods

The use of Etest for moulds has also shown excellent agreement with the broth microdilution method. In a comparison of 90 isolates, 100% agreement was seen with *Aspergillus* species and amphotericin and itraconazole, and >90% agreement with other species of moulds. In a large comparative study, 283 clinical isolates of *A. fumigatus* were tested for susceptibility to AP, IT, and VO with Etest and CLSI broth microdilution reference method. Agreement between the methods was 98–100% ±2 dilutions and 86–95% ±1 dilution after 48 h of incubation in RPMI medium [13]. Szekely et al. tested three isolates of *Aspergillus* that were resistant to itraconazole in an animal model and were associated with treatment failure had Etest MICs of >32 µg/ml. As has been shown for yeast testing, it was stated by the author that Etest was able to better discriminate in vitro amphotericin resistance in moulds compared to BMD [42]. When tested on RPMI 1640 media, better agreement was seen in a comparative study with Etest and 50 moulds with both RPMI and casitone agar and itraconazole [29]. A study of 146 clinical isolates of filamentous fungi (*Aspergillus* sp., *Mucor*, and *Rhizopus*) compared Etest to broth microdilution for posaconazole and had excellent categorical agreement (96–98%) [19].

Studies have also evaluated the use of Etest with other fungi such as dermatophytes, where a reference CLSI method is not published (Chap. 4).

3.3.4 Other Agar-Based Methods

Fluconazole testing has also been conducted based on comparison of colony size of the yeast with or without antifungal on yeast morphology agar. A study by Xu et al. compared fluconazole MIC values obtained from this method using ten strains of yeast in a comparison to the CLSI method. Colony size was measured microscopically using 50–100× magnification. The size of 20 random colonies on agar plates with 12 concentrations of FL was measured. The MIC was selected as the lowest concentration of drug that caused a significant decrease in colony size compared with the control [48]. Nine of the ten isolates tested had similar MICs with the two methods (±2 – twofold dilutions), and results were reproducible on repeat testing. This method could be a useful to screen for azole resistance.

A semisolid agar method using heart infusion broth, designed to mimic conditions in the body at the site of infection, e.g., low oxygen tension, uses 0.5% agar in tubes containing various interpretive breakpoint concentrations of the antifungal agent. The assay most closely resembled the format of a macrobroth dilution method. Growth in drug-containing tubes was read after 48 h of incubation and compared to growth in control tubes. The endpoint was considered the concentration at which there was 75% inhibition of growth as compared to the control [34]. Reproducible results were seen with QC yeast isolates, and 96% agreement was seen with clinical mould isolates.

3.4   Non-growth-Based Methods

Several methods of antifungal susceptibility testing have been evaluated using principles other than growth of the fungus in an attempt to provide more rapid results. RSA (rapid susceptibility assay) measures the uptake of glucose by the fungus, a process that will be suppressed when a susceptible fungus is exposed to that antifungal agent. The amount of residual glucose is measured in an automated reader after addition of an enzyme substrate bound to a colored compound. It is more rapid (6–19 h) than conventional methods [39].

3.4.1   Ergosterol Assay

Other unique methods of antifungal susceptibility testing include quantitation of ergosterol present in the fungal cell wall. This method is only useful for azoles, since they inhibit ergosterol synthesis. The procedure measures the total intracellular ergosterol content following growth of the organism in different concentrations of FL. The advantage of this assay is that it eliminates the subjectivity in reading endpoints associated with trailing of growth caused by azoles. Results for 18 FL susceptible *C. albicans* isolates, which did not show endpoint trailing, demonstrated 100% agreement with the CLSI method. However, poor agreement was seen for isolates that exhibited trailing of endpoints [3]. Isolates that were FL susceptible when read at 24 h yet resistant at 48 h by the broth microdilution were found to be FL susceptible by the ergosterol quantitation method. Another advantage of this method is that it does not require any specialized equipment and can be read after an 18-h incubation.

3.4.2   Flow Cytometry

Flow cytometry has been investigated as a rapid non-growth-based method for determining susceptibility of microorganisms to antimicrobial agents. This technique measures the change of fungal cell membrane potential, metabolic activity due to membrane damage or uptake of a DNA-binding dye in response to the addition of an antifungal agent [35]. DNA-binding dyes which have been used include acridine orange and FUN-1. FUN-1 is a fluorescent probe which is converted to cylindrical intravacuolar structures in metabolically active cells [5]. Good correlation was seen with the CLSI method for susceptibility of *Aspergillus* sp. to amphotericin. The main advantage of flow cytometry is that it does not require growth of the fungi; however, it does require a dedicated instrument, which is costly, and a highly trained technologist to perform the test. Most dyes used in flow cytometry assays measure death of the cells, and therefore fungicidal activity, while FUN-1 is a vital dye and measures both fungistatic and fungicidal effects.

3.4.3   Molecular Methods

Molecular methods for detection of antifungal resistance are being investigated in research laboratories in an effort to overcome some of the problems of phenotypic testing and to provide more rapid and specific results. However, resistance markers are not well established since resistance mechanisms for azoles are multifactorial and complex. The most common mechanisms of fluconazole resistance is either point mutations in erg11 gene that reduces binding of the drug to the target site or overexpression of drug efflux transporter genes above the level found in susceptible strains and encoded by the CDR1 and CDR2 genes [38]. In a study of 59 isolates of *Candida* species, resistant to fluconazole as tested by CLSI methods, all strains were found to have either the erg11 mutation or overexpression of efflux transporter genes [25].

Resistance to echinocandins resistance is due to mutations in FKS1 gene and is associated with elevated caspofungin MIC of >16 µg/ml and an increase in the amount of drug needed to decrease fungal colony counts in a mouse model. Balashov et al. used molecular beacons with real-time multiplex PCR to screen-resistant mutants of *C. albicans* for FKS1 gene mutations [6].

3.5   Beyond Susceptibility Testing

3.5.1   Minimum Fungicidal Concentration (MFC)

Fungicidal testing can be useful in certain clinical situations and most notably in serious infections in immunocompromised patients. Since most serious fungal infections occur in the compromised host, fungicidal tests may be needed to further optimize the management of antifungal therapy in this patient group. Assays of the bactericidal activity of antibiotics include time-kill studies and the determination of the minimum bactericidal concentration (MBC) based on 99.9% kill of the initial inoculum in terms of CFU/ml. Technical parameters that affect these assays for both bacteria and fungi include: inoculum density, growth phase, drug carryover, test medium, test format (macro versus micro), method of sampling the growth, trailing of growth, and paradoxical effects of drugs. Measurement of dose-related mortality in animal models would be the gold standard for fungicidal testing; however these studies are not routinely done. The advantage of the time-kill assay is that it can assess the rate and extent of killing [32]. Although there are no CLSI guidelines, in vitro and animal data suggest a better correlation between time-kill data rather than MIC values with clinical outcomes in patients with *Aspergillus* infection, particularly those with *A. terreus* infection and treated with amphotericin.

3.5.2 Antifungal Combination Testing

Combinations of antifungal drugs are often used for the treatment of serious fungal infections associated with high morbidity and mortality especially in the compromised host. Amphotericin B or fluconazole are commonly used in combination with flucytosine. However, combination testing is rarely performed in the clinical laboratory, in part due to the difficulty of setting up the assay and in part because no defined standards exist for the procedure to be used and interpretation of the result [41]. Data in the literature suggest in vitro antagonism between amphotericin B and azoles, although clinical trials have not confirmed these findings [36]. Current methods for combination testing of antibacterial agents include checkerboard titration, time-kill studies, and Etest methods. All of these methods have been evaluated for use with fungi [21]. In a study comparing the three methods with three strains of *Candida*, Etest and time-kill had the best agreement [16].

3.6 Summary

Routine testing of yeast, particularly *Candida* sp. not albicans from sterile sites, is currently recommended by CLSI as well as experts in the field [37]. With the advent of FDA-cleared user-friendly methods, "real-time" testing is appropriate for routine clinical laboratories. Testing of yeast isolates to create a yearly antibiogram is also recommended to aid in choice of empiric therapy. Routine testing of moulds is not recommended at this time; however, testing of isolates in select clinical situations may provide useful information for choice of antifungal therapy for serious infections.

References

1. Alexander BD, Byrne TC, Smith KL, Hanson KE, Anstrom KJ, Perfect JR, Reller LB (2007) Comparative evaluation of Etest and sensititre yeastone panels against the Clinical and Laboratory Standards Institute M27-A2 reference broth microdilution method for testing Candida susceptibility to seven antifungal agents. J Clin Microbiol 45:698–706
2. Antachopoulos C, Meletiadis J, Roilides E, Sein T, Walsh TJ (2006) Rapid susceptibility testing of medically important zygomycetes by XTT assay. J Clin Microbiol 44:553–560
3. Arthington-Skaggs BA, Jradi H, Desai T, Morrison CJ (1999) Quantitation of ergosterol content: novel method for determination of fluconazole susceptibility of Candida albicans. J Clin Microbiol 37:3332–3337
4. Balajee SA, Imhof A, Gribskov JL, Marr KA (2005) Determination of antifungal drug susceptibilities of Aspergillus species by a fluorescence-based microplate assay. J Antimicrob Chemother 55:102–105
5. Balajee SA, Marr KA (2002) Conidial viability assay for rapid susceptibility testing of Aspergillus species. J Clin Microbiol 40:2741–2745

6. Balashov SV, Park S, Perlin DS (2006) Assessing resistance to the echinocandin antifungal drug caspofungin in Candida albicans by profiling mutations in FKS1. Antimicrob Agents Chemother 50:2058–2063
7. Barry AL, Pfaller MA, Rennie RP, Fuchs PC, Brown SD (2002) Precision and accuracy of fluconazole susceptibility testing by broth microdilution, Etest, and disk diffusion methods. Antimicrob Agents Chemother 46:1781–1784
8. Bolmström A, Arvidson S, Ericsson M, Karlsson A (1988) A Novel Technique for direct quantification of antimicrobial susceptibility of microorganisms. Poster 1209 ICAAC, Los Angeles
9. Cuenca-Estrella M, Moore CB, Barchiesi F, Bille J, Chryssanthou E, Denning DW, Donnelly JP, Dromer F, Dupont B, Rex JH, Richardson MD, Sancak B, Verweij PE, Rodriguez-Tudela JL (2003) Multicenter evaluation of the reproducibility of the proposed antifungal susceptibility testing method for fermentative yeasts of the Antifungal Susceptibility Testing Subcommittee of the European Committee on Antimicrobial Susceptibility Testing (AFST-EUCAST). Clin Microbiol Infect 9:467–474
10. Davey KG, Holmes AD, Johnson EM, Szekely A, Warnock DW (1998) Comparative evaluation of FUNGITEST and broth microdilution methods for antifungal drug susceptibility testing of Candida species and Cryptococcus neoformans. J Clin Microbiol 36:926–930
11. Espinel-Ingroff A (2006) Comparison of three commercial assays and a modified disk diffusion assay with two broth microdilution reference assays for testing zygomycetes, Aspergillus spp., Candida spp., and Cryptococcus neoformans with posaconazole and amphotericin B. J Clin Microbiol 44:3616–3622
12. Espinel-Ingroff A, Pfaller M, Messer SA, Knapp CC, Killian S, Norris HA, Ghannoum MA (1999) Multicenter comparison of the sensititre YeastOne Colorimetric Antifungal Panel with the National Committee for Clinical Laboratory Standards M27-A reference method for testing clinical isolates of common and emerging Candida spp., Cryptococcus spp., and other yeasts and yeast-like organisms. J Clin Microbiol 37:591–595
13. Guinea J, Pelaez T, Alcala L, Bouza E (2007) Correlation between the Etest and the CLSI M-38 A microdilution method to determine the activity of amphotericin B, voriconazole, and itraconazole against clinical isolates of Aspergillus fumigatus. Diagn Microbiol Infect Dis 57:273–276
14. Imhof A, Balajee SA, Marr KA (2003) New methods to assess susceptibilities of Aspergillus isolates to caspofungin. J Clin Microbiol 41:5683–5688
15. Kartsonis N, Killar J, Mixson L, Hoe CM, Sable C, Bartizal K, Motyl M (2005) Caspofungin susceptibility testing of isolates from patients with esophageal candidiasis or invasive candidiasis: relationship of MIC to treatment outcome. Antimicrob Agent Chemother 49:3616–3623
16. Lewis RE, Diekema DJ, Messer SA, Pfaller MA, Klepser ME (2002) Comparison of Etest, chequerboard dilution and time-kill studies for the detection of synergy or antagonism between antifungal agents tested against Candida species. J Antimicrob Chemother 49:345–351
17. Liao RS, Rennie RP, Talbot JA (2001) Novel fluorescent broth microdilution method for fluconazole susceptibility testing of Candida albicans. J Clin Microbiol 39:2708–2712
18. Maxwell MJ, Messer SA, Hollis RJ, Diekema DJ, Pfaller MA (2003) Evaluation of Etest method for determining voriconazole and amphotericin B MICs for 162 clinical isolates of Cryptococcus neoformans. J Clin Microbiol 41:97–99
19. Messer SA, Diekema DJ, Hollis RJ, Boyken LB, Tendolkar S, Kroeger J, Pfaller MA (2007) Evaluation of disk diffusion and Etest compared to broth microdilution antifungal susceptibility testing of posaconazole against clinical isolates of filamentous fungi. J Clin Microbiol 45:1322–1324
20. Morace G, Amato G, Bistoni F, Fadda G, Marone P, Montagna MT, Oliveri S, Polonelli L, Rigoli R, Mancuso I, La Face S, Masucci L, Romano L, Napoli C, Tato D, Buscema MG, Belli CMC, Piccirillo MM, Conti S, Covan S, Fanti F, Cavanna C, D'Alo F, Pitzurra L (2002) Multicenter comparative evaluation of six commercial systems and the National Committee for Clinical Laboratory Standards M27-A broth microdilution method for fluconazole susceptibility testing of Candida species. J Clin Microbiol 40:2953–2958

21. Mukherjee PK, Sheehan DJ, Hitchcock CA, Ghannoum MA (2005) Combination treatment of invasive fungal infections. Clin Microbiol Rev 18:163–194
22. NCCLS (2004) Reference method for antifungal disk diffusion susceptibility testing of yeasts, approved standard. NCCLS, Wayne
23. NCCLS (2002) Reference method for broth dilution antifungal susceptibility testing of filamentous fungi, approved standard. NCCLS, Wayne
24. NCCLS (2002) Reference method for broth dilution antifungal susceptibility testing of yeasts. Approved standard, 2nd edn. NCCLS, Wayne
25. Park S, Perlin DS (2005) Establishing surrogate markers for fluconazole resistance in Candida albicans. Microb Drug Resist 11:232–238
26. Pfaller MA, Boyken L, Hollis RJ, Messer SA, Tendolkar S, Diekema DJ (2004) Clinical evaluation of a dried commercially prepared microdilution panel for antifungal susceptibility testing of five antifungal agents against Candida spp. and Cryptococcus neoformans. Diagn Microbiol Infect Dis 50:113–117
27. Pfaller MA, Diekema DJ, Procop GW, Rinaldi MG (2007) Multicenter comparison of the VITEK 2 yeast susceptibility test with the CLSI broth microdilution reference method for testing fluconazole against Candida spp. J Clin Microbiol 45:796–802
28. Pfaller MA, Messer SA, Coffmann S (1995) Comparison of visual and spectrophotometric methods of MIC endpoint determinations by using broth microdilution methods to test five antifungal agents, including the new triazole D0870. J Clin Microbiol 33:1094–1097
29. Pfaller MA, Messer SA, Mills K, Bolmstrom A (2000) In vitro susceptibility testing of filamentous fungi: comparison of Etest and reference microdilution methods for determining itraconazole MICs. J Clin Microbiol 38:3359–3361
30. Pfaller MA, Messer SA, Mills K, Bolmstrom A, Jones RN (2001) Evaluation of Etest method for determining caspofungin (MK-0991) susceptibilities of 726 clinical isolates of Candida species. J Clin Microbiol 39:4387–4389
31. Pfaller MA, Messer SA, Mills K, Bolmstrom A, Jones RN (2001) Evaluation of Etest method for determining posaconazole MICs for 314 clinical isolates of Candida species. J Clin Microbiol 39:3952–3954
32. Pfaller MA, Sheehan DJ, Rex JH (2004) Determination of fungicidal activities against yeasts and molds: lessons learned from bactericidal testing and the need for standardization. Clin Microbiol Rev 17:268–280
33. Philpot C (1993) Determination of sensitivity to antifungal drugs: evaluation of an API kit. Br J Biomed Sci 50:27–30
34. Provine H, Hadley S (2000) Preliminary evaluation of a semisolid agar antifungal susceptibility test for yeasts and molds. J Clin Microbiol 38:537–541
35. Ramani R, Chaturvedi V (2000) Flow cytometry antifungal susceptibility testing of pathogenic yeasts other than Candida albicans and comparison with the NCCLS broth microdilution test. Antimicrob Agents Chemother 44:2752–2758
36. Rex JH, Pappas PG, Karchmer AW, Sobel J, Edwards JE, Hadley S, Brass C, Vazquez JA, Chapman SW, Horowitz HW, Zervos M, McKinsey D, Lee J, Babinchak T, Bradsher RW, Cleary JD, Cohen DM, Danziger L, Goldman M, Goodman J, Hilton E, Hyslop NE, Kett DH, Lutz J, Rubin RH, Scheld WM, Schuster M, Simmons B, Stein DK, Washburn RG, Mautner L, Chu TC, Panzer H, Rosenstein RB, Booth J (2003) A randomized and blinded multicenter trial of high-dose fluconazole plus placebo versus fluconazole plus amphotericin B as therapy for candidemia and its consequences in nonneutropenic subjects. Clin Infect Dis 36:1221–1228
37. Rex JH, Pfaller MA, Walsh TJ, Chaturvedi V, Espinel-Ingroff A, Ghannoum MA, Gosey LL, Odds FC, Rinaldi MG, Sheehan DJ, Warnock DW (2001) Antifungal susceptibility testing: practical aspects and current challenges. Clin Microbiol Rev 14:643–658, table of contents
38. Ribeiro MA, Paula CR (2007) Up-regulation of ERG11 gene among fluconazole-resistant Candida albicans generated in vitro: is there any clinical implication? Diagn Microbiol Infect Dis 57:71–75
39. Riesselman MH, Hazen KC, Cutler JE (2000) Determination of antifungal MICs by a rapid susceptibility assay. J Clin Microbiol 38:333–340

40. Sims CR, Paetznick VL, Rodriguez JR, Chen E, Ostrosky-Zeichner L (2006) Correlation between microdilution, E-test, and disk diffusion methods for antifungal susceptibility testing of posaconazole against Candida spp. J Clin Microbiol 44:2105–2108
41. Steinbach WJ, Stevens DA, Denning DW (2003) Combination and sequential antifungal therapy for invasive aspergillosis: review of published in vitro and in vivo interactions and 6281 clinical cases from 1966 to 2001. Clin Infect Dis 37(Suppl 3):S188–S224
42. Szekely A, Johnson EM, Warnock DW (1999) Comparison of E-test and broth microdilution methods for antifungal drug susceptibility testing of molds. J Clin Microbiol 37:1480–1483
43. Tan GL, Peterson EM (2005) CHROMagar Candida medium for direct susceptibility testing of yeast from blood cultures. J Clin Microbiol 43:1727–1731
44. van Eldere J, Joosten L, Verhaeghe V, Surmont I (1996) Fluconazole and amphotericin B antifungal susceptibility testing by National Committee for Clinical Laboratory Standards broth macrodilution method compared with E-test and semiautomated broth microdilution test. J Clin Microbiol 34:842–847
45. Vandenbossche I, Vaneechoutte M, Vandevenne M, De Baere T, Verschraegen G (2002) Susceptibility testing of fluconazole by the NCCLS broth macrodilution method, E-Test, and disk diffusion for application in the routine laboratory. J Clin Microbiol 40:918–921
46. Wanger A, Mills K, Nelson P, Rex J (1995) Comparison of Etest and National Committee for Clinical Laboratory Standards broth macrodilution method for antifungal susceptibility testing: enhanced ability to detect amphotericin B-resistant Candida isolates. Antimicrob Agents Chemother 39:2520–2522
47. Willinger B, Engelmann E, Hofmann H, Metzger S, Apfalter P, Hirschl AM, Makristathis A, Rotter M, Raddatz B, Seibold M (2002) Multicenter comparison of Fungitest for susceptibility testing of Candida species. Diagn Microbiol Infect Dis 44:253–257
48. Xu J, Vilgalys R, Mitchell TG (1998) Colony size can be used to determine the MIC of fluconazole for pathogenic yeasts. J Clin Microbiol 36:2383–2385
49. Zambardi G, Parreno D, Monnin V, Fothergill A, Hurt L, Bassel A, McCarthy D, Canard I, Slaughter J (2005) Presented at the interscience conference on antimicrobial agents and chemotherapy, Washington, DC. Rapid antifungal susceptibility testing of medically important yeasts with the VITEK 2 system

Chapter 4
Susceptibility Testing of Dermatophytes

David V. Chand and Mahmoud A. Ghannoum

Abstract The dermatophytes are a specialized group of fungi which infect the keratinized tissues of humans (hair, nails, and skin) and cause superficial infections. While several studies have been conducted to develop methods to determine the susceptibilities of yeast and filamentous fungi, similar studies for dermatophytes have only recently taken place. In this chapter, we will review how susceptibility testing of dermatophytes was developed and how it has already been applied to clinical samples. With several agents now available for treating infections due to dermatophytes, susceptibility testing will serve as a valuable tool for clinicians as they choose the most appropriate treatment option. Studies are still needed to establish interpretive breakpoints for antifungal agents used in the treatment of superficial fungal infections.

4.1 Introduction

4.1.1 Background

The dermatophytes are a specialized group of fungi which infect the keratinized tissues of humans, such as skin, hair, and nails, commonly causing superficial infections. While there have been several studies involving yeasts which have been used to develop the Clinical and Laboratory Standards Institute (CLSI) reference method

D.V. Chand, MSE, M.D.
Division of Pediatric Infectious Diseases and Rheumatology, Department of Pediatrics,
Rainbow Babies & Children's Hospital/University Hospitals of Cleveland,
Cleveland, OH, USA

M.A. Ghannoum, M.Sc., Ph.D. (✉)
Center for Medical Mycology, University Hospitals of Cleveland/Case
Western Reserve University, Cleveland, OH, USA
e-mail: mag3@cwru.edu

G.S. Hall (ed.), *Interactions of Yeasts, Moulds, and Antifungal Agents:*
How to Detect Resistance, DOI 10.1007/978-1-59745-134-5_4,
© Springer Science+Business Media, LLC 2012

[1], the antifungal susceptibility of dermatophytes has only recently been addressed in the standard for filamentous fungi [2]. While some infections do respond to topical therapy, others, particularly involving the scalp and nails, require prolonged systemic therapy. Therefore, determining the susceptibility patterns of dermatophytes will allow clinicians to choose the most appropriate antifungal therapy.

4.1.2 Dermatophytes and Clinical Manifestations

For an extensive review of this topic [3], please refer to the reference section. Briefly, the dermatophytes which cause human disease include *Epidermophyton*, *Microsporum*, and *Trichophyton* species. Among the fungi, they are unique in their ability to cause communicable disease, as infections can result from contact with humans, animals, and fomites. Common clinical manifestations include: (1) tinea capitis (scalp); (2) tinea barbae (bearded area); (3) tinea corporis (trunk, limbs, face); (4) tinea cruris (groin, perianal, and perineal areas); (5) tinea pedis (feet); and (6) tinea unguium, often called onychomycosis (nails).

4.1.3 Need for Susceptibility Testing

While the number of fungal infections is increasing, so is the population of immunocompromised hosts, where such infections can be more extensive and difficult to treat. Furthermore, there are several antifungal agents available to treat infections due to dermatophytes, including ciclopirox, fluconazole, griseofulvin, itraconazole, posaconazole, terbinafine, and voriconazole. Of these, only griseofulvin, ciclopirox, terbinafine, and itraconazole are approved by the United States Food and Drug Administration (FDA) to treat superficial infections. The other agents showed promising in vitro data against dermatophytes [4]. The availability of susceptibility patterns will help guide the management of such infections and provide the opportunity for the clinician to select the therapeutic option that maximizes efficacy, safety, and convenience while minimizing cost and toxicity [5, 6]. In addition, susceptibility testing will provide a means to monitor the development of resistance and predict the therapeutic potential of investigational agents.

4.2 Susceptibility Testing

4.2.1 Optimal Growth Conditions

Both the CLSI M27-A3 and M38-A2 reference methods specify inoculum size and preparation, test medium, incubation time and temperature, and end-point definitions for determining the antifungal susceptibility against yeast and filamentous fungi, respectively. Before susceptibility testing of dermatophytes could be standardized, the optimal growth conditions for different dermatophytes needed to be determined. Norris et al. [7] studied 18 clinical specimens of *Trichophyton* species (*T. rubrum, T. mentagrophytes, T. tonsurans*) isolated from nail or hair. Four types of media were examined: RPMI 1640 with L-glutamine, without sodium bicarbonate and buffered at pH = 7.0; antibiotic medium #3 (Penassay); yeast nitrogen base with 0.5% dextrose buffered at pH = 7.0; and Sabouraud dextrose broth. Growth was evaluated at 30°C and 35°C. After the optimal media and temperature were determined, the effect of inoculum concentration (10^3, 10^4, and 10^5 conidia/mL) against four antifungal agents (griseofulvin, itraconazole, fluconazole, and terbinafine) was studied.

RPMI 1640 and Sabouraud dextrose broth both supported optimal growth. The authors concluded that RPMI 1640 was the best choice because it is a chemically defined medium with no known interference with antifungal agents. It was also found that 4 days provided sufficient time for adequate growth and, from a practical sense, a reasonable time to provide clinicians with the necessary data upon which to guide their management. No difference in growth was found at the two temperatures, but the authors added that 35°C was a convenient temperature because additional plates for yeast testing could be incubated concurrently. The minimum inhibitory concentration (MIC) was defined as the point at which the organism was inhibited by 80% as compared with growth in the control well. Inoculum size did not affect the MIC results for itraconazole or terbinafine, but the larger inoculum sizes did result in a slightly higher MIC for griseofulvin and a significantly higher MIC for fluconazole. Therefore, 10^3 conidia/mL was determined to be the optimal inoculum size.

Jessup et al. [8] furthered this study by determining the optimal medium for conidial formation by dermatophytes. The identification of such a medium is critical because dermatophytes, especially *T. rubrum*, are poor producers of conidia, and failure to produce spores will limit one's ability to prepare the proper inoculum size needed for testing. Both oatmeal cereal agar and rice agar best supported the ability of *T. rubrum* to produce conidia, in contrast to potato dextrose and Mycosel with 1% yeast extract. Nevertheless, 15% of *T. rubrum* isolates still failed to produce any conidia in the oatmeal cereal and rice agar media.

In summary, establishing the optimal growth conditions was the first step in developing a method to determine the antifungal susceptibilities of dermatophytes. Oatmeal cereal agar should be used to promote conidia formation, particularly for *T. rubrum*, so that an initial inoculum size of 10^3 conidia/mL can be measured. RPMI 1640 should then be used as the growth medium, and incubation should occur for 4 days at 35°C. The MIC is defined as the point at which the organism is inhibited 80% compared with the growth in the control sample.

4.2.2 Standardization

Demonstrating reproducibility of endpoints and the ability to detect the development of resistance is an integral part of establishing a susceptibility testing method (inter- and intralaboratory agreement). A multicenter study involving six laboratories examined the MIC reproducibility of seven antifungal agents tested against 25 dermatophyte

Table 4.1 Interlaboratory agreement summary

Antifungal	Within 1 dilution	Within 2 dilutions	Within 3 dilutions
	% of total isolates		
50% Inhibition endpoint			
Ciclopirox	94.6	96.0	99.0
Fluconazole	67.6	82.1	93.6
Griseofulvin	96.9	98.6	98.6
Itraconazole	63.9	79.8	92.0
Posaconazole	93.6	99.4	100
Terbinafine	84.5	92.9	98.0
Voriconazole	75.3	87.8	100
80% Inhibition endpoint			
Ciclopirox	95.9	97.6	99.0
Fluconazole	64.9	85.8	87.8
Griseofulvin	88.9	97.7	98.4
Itraconazole	62.8	82.7	90.5
Posaconazole	88.8	95.6	99.3
Terbinafine	85.8	95.3	98.0
Voriconazole	73.6	89.1	97.5

Reprinted from Ref. [4]. With permission from the American Society for Microbiology

Table 4.2 Intralaboratory agreement summary

Antifungal	Within 1 dilution	Within 2 dilutions	Within 3 dilutions
	% of total isolates		
50% Inhibition endpoint			
Ciclopirox	91.8	96.6	99.3
Fluconazole	79.6	87.8	93.9
Griseofulvin	91.2	97.3	98.0
Itraconazole	73.5	86.4	95.2
Posaconazole	95.2	98.6	100
Terbinafine	84.4	91.2	97.3
Voriconazole	74.1	86.3	94.5
80% Inhibition endpoint			
Ciclopirox	92.5	100	100
Fluconazole	74.8	83.6	89.0
Griseofulvin	91.8	96.6	98.3
Itraconazole	66.7	85.1	91.9
Posaconazole	89.8	97.3	100
Terbinafine	83.7	91.2	96.0
Voriconazole	74.1	87.4	96.2

Reprinted from Ref. [4]. With permission from the American Society for Microbiology

isolates using the susceptibility method outlined in the previous section [4]. Each laboratory tested five blinded pairs of five dermatophyte strains: *T. rubrum, T. mentagrophytes, T. tonsurans, E. floccosum, and M. canis*. The antifungal agents included ciclopirox, fluconazole, griseofulvin, itraconazole, posaconazole, terbinafine, and voriconazole.

Table 4.3 Quality control ranges for antifungal agents

Dermatophyte QC	Antifungal agent	MIC range (µg/mL)
T. mentagrophytes ATCC MYA-4439	Ciclopirox	0.5–2.0
	Griseofulvin	0.12–0.5
	Itraconazole	0.03–0.25
	Posaconazole	0.03–0.25
	Terbinafine	0.002–0.008
	Voriconazole	0.03–0.25
T. rubrum ATCC MYA-4438	Ciclopirox	0.5–2.0
	Fluconazole	0.5–4.0
	Voriconazole	0.008–0.06

Reprinted from Ref. [2]

The results can be seen in Tables 4.1 and 4.2. Using 80% inhibition as compared to the growth control as the endpoint, interlaboratory agreement within three dilutions ranged from 87.8% for fluconazole to 99.3% for posaconazole. Similarly, the intralaboratory agreement within three dilutions ranged from 89% for fluconazole to 100% for ciclopirox and posaconazole.

In order for this method to be approved by CLSI and to have utility in clinical microbiology laboratories, appropriate quality control/standard isolates needed to be identified. Eight laboratories tested a total of ten different dermatophyte strains against the seven antifungal agents examined in the interlaboratory study previously described [9]. The candidate strains included five *T. rubrum* strains, known to have elevated MICs to terbinafine, as well as five strains of *T. mentagrophytes*. Based on best overall agreement, *T. mentagrophytes* ATCC MYA-4439 and *T. rubrum* ATCC MYA-4438 were included as reference strains (Table 4.3).

Before having a clinical utility, it will be important to develop interpretive breakpoints for different antifungal agents. These breakpoints will guide the clinicians in deciding whether an organism is susceptible or resistant to a particular agent. Further work in this area is warranted.

4.2.3   Clinical Applications

Although breakpoints have not yet been established, Bradley et al. used antifungal susceptibility testing to investigate 104 *T. rubrum* isolates from 30 patients who had failed treatment with terbinafine for onychomycosis of the toenail [10]. There was no increase in MIC of sequential isolates seen and, therefore, no development of resistance. The authors concluded that treatment failure was likely due to host factors.

Another study was conducted to determine the prevalence of dermatophyte-positive scalp cultures among elementary school students in Cleveland, Ohio [11]. The MIC values of griseofulvin, itraconazole, fluconazole, and terbinafine against the 122 isolates obtained were measured. The MIC_{50}, the lowest concentration at which 50% of all isolates were inhibited, was 2.0, 0.25, 0.004, and 0.015 µg/mL for fluconazole, griseofulvin, itraconazole, and terbinafine, respectively. Similarly, the MIC_{90}, the lowest concentration at which 90% of all isolates were inhibited, was 8.0, 1.0, 0.03, and 0.03 µg/mL, respectively. The MICs were found to be well below achievable skin levels of most of the drugs tested.

Finally, a 12-center North American study was conducted to determine the frequency of onychomycosis, identify the responsible pathogens, and determine the antifungal susceptibility of the isolates [12]. Of the 1,832 participants in the study, 253 (13.8%) met the case definition for onychomycosis. Dermatophytes were the most commonly isolated fungi (59%) from all nail samples. The MICs of terbinafine, griseofulvin, fluconazole, and itraconazole were measured, and all the dermatophyte isolates appeared to be susceptible. Nevertheless, terbinafine appeared to possess the greatest activity with much smaller MICs.

4.3 Conclusions

Although susceptibility testing for fungi has lagged behind that for bacteria, much progress has been made over the last decade. Such knowledge should improve the clinician's ability to select the best choice among the growing arsenal of antifungal agents available. Further work needs to be done to correlate in vitro findings with clinical outcomes. The collection of such data will allow for the establishment of interpretive breakpoints, which have already been established for *Candida* species and some antifungal agents (e.g., anidulafungin, caspofungin, fluconazole, 5-fluorocytosine, itraconazole, micafungin, and voriconazole). Finally, surveillance studies are needed to determine the true frequency of antifungal resistance. The development of the method for susceptibility testing of dermatophytes is a step in the right direction toward reaching these goals.

References

1. CLSI (2008) Reference method for broth dilution antifungal susceptibility testing of yeasts; approved standard – third edition. CLSI document M27-A3. Clinical and Laboratory Standards Institute, CLSI, 940 West Valley Road, Suite 1400, Wayne, PA 19087–1898 USA
2. CLSI (2008) Reference method for broth dilution antifungal susceptibility testing of filamentous fungi; approved standard- second edition. CLSI document M38-A2 [ISBN 1–56238–668–9]. CLSI, 940 West Valley Road, Suite 1400, Wayne, PA 19087–1898 USA
3. Weitzman I, Summerbell RC (1995) The dermatophytes. Clin Microbiol Rev 8(2):240–259
4. Ghannoum MA, Chaturvedi V, Espinel-Ingroff A et al (2004) Intra- and interlaboratory study of a method for testing the antifungal susceptibilities of dermatophytes. J Clin Microbiol 42(7):2977–2979
5. Ghannoum MA, Rex JH, Galgiani JN (1996) Susceptibility testing of fungi: current status and correlation of *in vitro* data with clinical outcome. J Clin Microbiol 34(3):489–495

6. Rex JH, Pfaller MA, Walsh TJ et al (2001) Antifungal susceptibility testing: practical aspects and current challenges. Clin Microbiol Rev 14(4):643–658

7. Norris HA, Elewski BE, Ghannoum MA (1999) Optimal growth conditions for the determination of the antifungal susceptibility of three species of dermatophytes with the use of a microdilution method. J Am Acad Dermatol 40(6):S9–S13

8. Jessup CJ, Warner J, Isham N, Hasan I, Ghannoum MA (2000) Antifungal susceptibility testing of dermatophytes: establishing a medium for inducing conidial growth and evaluation of susceptibility of clinical isolates. J Clin Microbiol 38(1):341–344

9. Ghannoum MA, Chaturvedi V, Espinel-Ingroff A et al. (submitted) An interlaboratory QC guidelines study for a proposed antifungal susceptibility method for the testing of dermatophytes, modified from CLSI M38-A. J Clin Microbiol

10. Bradley MC, Leidich S, Isham N, Elewski BE, Ghannoum MA (1999) Antifungal susceptibilities and genetic relatedness of serial Trichophyton rubrum isolates from patients with onychomycosis of the toenail. Mycoses 42(suppl 2):105–110

11. Ghannoum MA, Isham N, Hajjeh R et al (2003) Tinea capitis in Cleveland: survey of elementary school students. J Am Acad Dermatol 48:189–193

12. Ghannoum MA, Hajjeh RA, Scher R et al (2000) A large-scale North American study of fungal isolates from nails: frequency of onychomycosis, fungal distribution, and antifungal susceptibility patterns. J Am Acad Dermatol 43:641–648

Chapter 5
Usual Susceptibility Patterns of Common Yeasts

Gerri S. Hall

Abstract The in vitro susceptibility patterns of some of the common yeast such as *Candida* spp., *Cryptococcus* spp., *Rhodotorula* spp., *Trichosporon* spp., *Saccharomyces cerevisiae* are described in this chapter. Isolates such as *C. albicans* and *C. parapsilosis* are fairly predictable; others such as *C. glabrata* may demonstrate less predictable patterns that require that susceptibility be performed for each isolate of significance. In this chapter, we describe the most common susceptibility patterns as found in the published literature, including some geographic differences throughout various parts of the world where results have been reported. Table 5.1 provides a summary of the expected susceptibility patterns for the most common yeast isolates described in this chapter.

5.1 *Candida* spp.

5.1.1 *C. albicans*

Most *Candida albicans* are susceptible in vitro to amphotericin-B, the azoles, fluconazole, itraconazole, voriconazole, and posaconazole, to 5-FC and to the echinocandins. This is especially true if the patient, from which they are isolated, has not be treated with any of these antifungals before the yeast is first isolated and the susceptibility test is performed. The ARTEMIS Global Surveillance Study data analyzing trending for an 8.5-year period of time with thousands of *C. albicans* isolates

G.S. Hall, Ph.D. (✉)
Section of Clinical Microbiology, Department of Clinical Pathology,
Cleveland Clinic, Cleveland, OH 44195, USA
e-mail: hallg@ccf.org

G.S. Hall (ed.), *Interactions of Yeasts, Moulds, and Antifungal Agents:*
How to Detect Resistance, DOI 10.1007/978-1-59745-134-5_5,
© Springer Science+Business Media, LLC 2012

worldwide could not demonstrate any more than 1.6% resistance to fluconazole and voriconazole from 1997 through 2005. There was slightly more resistance found in the >4,600 isolates from North America, with 5% fluconazole resistance reported and 3.7% voriconazole resistance [20]. The SENTRY study results for 2006–2007 in North America, Europe, and Latin America were similar in that 771 isolates of *C. albicans* demonstrated overall 99.7% susceptibility and 0.3% S-dose dependent (DD) to fluconazole and 100% susceptibility to voriconazole; in isolates from North America alone, there was a 99.5% susceptibility – so a very small percentage of non-susceptible isolates were found. In the latter study, there was 100% susceptibility to amphotericin-B, caspofungin, and anidulafungin; 99.7% and 97.7% susceptibility to 5-FC and itraconazole were seen, respectively. The Minimum Inhibitory Concentration ($MIC_{90\%}$) for posaconazole was 0.12 µg/ml [17]. Susceptibility testing for *C. albicans* probably can be reserved for instances where failures have occurred or in patients for whom azoles have been used in prophylaxis or treatment of other fungal organisms.

5.1.2   C. glabrata

C. glabrata is second only to *Candida albicans* in frequency of isolation of *Candida* spp. from clinical samples, in particular, blood, urine, and vaginal samples. It is a more resistant isolate than is *C. albicans*, especially to the agent fluconazole. In a recent ARTEMIS Surveillance Study report of >23,000 clinical *C. glabrata* isolates from five geographic regions in the world, decreased susceptibility to fluconazole occurred throughout all regions, from 63% to 77% susceptibility depending upon the region. Poland, Czech Republic, Venezuela, and Greece had the highest overall rates of resistance to fluconazole. In the USA, the susceptibility was 74–92%. Voriconazole activity was superior to that of fluconazole throughout all regions [23]. There is significant variability, hence susceptibility testing, dependent upon clinical significance of the *C. glabrata* isolation, should be considered before employing an azole for therapy. In an 8.5-year analysis of global trends in fluconazole susceptibility, between 1997 and 2005, resistance had decreased from a high of 19–15% in 2005. For voriconazole, the percent resistance has remained fairly stable at just under 10%. Rates of resistance globally are highest for both fluconazole and voriconazole versus *C. glabrata* in North America, however [20]. Data from the SENTRY Surveillance studies reported that 100% of 202 isolates of *C. glabrata* from North America, Latin America, and Europe remained fully susceptible to amphotericin-B, 5-FC, caspofungin, and anidulafungin between 2006 and 2007; 74% of the isolates were susceptible, 15% S-DD, and 10% resistant to fluconazole versus 90% susceptible to voriconazole. Itraconazole performed poorly, with 69.8% of the *C. glabrata* isolates falling in the resistant range [17]. Rates of resistance to the azoles for C. *glabrata* are variable between regions of the world and from 1 year to the next, especially in regard to fluconazole. Because of the potential lack of predictability of *C. glabrata*, these results would suggest that susceptibility testing be performed when consideration is given to the use of an azole, especially fluconazole, for clinically significant infections with *C. glabrata*.

5.1.3   C. parapsilosis

Candida parapsilosis is usually seen as one of the top four non-*C. albicans* species of *Candida* isolated from clinical specimens, although this frequency varies throughout the world. Global susceptibility data versus fluconazole and voriconazole has been reported from the ARTEMIS surveillance studies published in 2008 of nearly 9400 isolates of *C. parapsilosis* in many regions of the world. Most regions demonstrated a high level of susceptibility to fluconazole and voriconazole (95% S and 98% S, respectively); however, rates of non-susceptibility in Africa and the Middle East were as high as 20% and 14% to fluconazole and voriconazole, respectively. Over 99% of >1,400 isolates of *C. parapsilosis* were found susceptible to caspofungin, with similar high levels of susceptibility to micafungin and anidulafungin as well [22]. When a trend analysis of resistance to fluconazole and voriconazole was reported for 1997–2005 worldwide, the rate of fluconazole resistance remained stable at ≤4.2%; for voriconazole ≤2.3% [20]. All but 3.4% of 238 isolates of *C. parapsilosis* in the SENTRY surveillance report of testing globally in 2006–2007 were susceptible to fluconazole and 0.4% non-susceptible to voriconazole. On the contrary, itraconazole performed poorly, with as many as 57% of the isolates testing as S-DD. There were rare isolates of *C. parapsilosis* that were non-susceptible to the echinocandins, slightly higher for anidulafungin than caspofungin. In addition, only 0.4% of the 238 isolates were found resistant to amphotericin and 1.3% resistant to 5-FC [17].

5.1.4   C. tropicalis

Candida tropicalis remains susceptible to amphotericin-B, azoles, and the echinocandins. The ARTEMIS global susceptibility report in 2007 showed that <6.6% of *C. tropicalis* isolates were resistant to fluconazole between 1997 and 2005, and the rates were fairly stable over these years, with the highest rates (~8%) in Asia-Pacific. The resistance rates for voriconazole were <5% for most of the years tabulated, although rates reached 8.1% in 2002 [20]. The SENTRY study reported that of 157 isolates tested during 2006–2007 worldwide, none were resistant to caspofungin, anidulafungin, or fluconazole. There was a 1.9% resistance seen versus amphotericin-B; 1.3% resistance was seen to voriconazole; and 34% of the 157 isolates were non-susceptible to itraconazole. Isolates of *C. tropicalis* usually do demonstrate lower susceptibility among the *Candida* spp. to 5-FC and in this study, the resistance was found to be 4.5%.

5.1.5   *C. krusei*

Candida krusei continues to be less frequently isolated than *C. glabrata, C. tropicalis*, and *C. parapsilosis* among the non–*C. albicans Candida* spp. In one global study, the highest incidence of finding *C. krusei* was seen in the Czech Republic and the lowest rates of recovery in Indonesia, South Korea, and Thailand. All isolates of *C. krusei* are considered clinically resistant to fluconazole. Susceptibility to voriconazole in the ARTEMIS global surveillance studies, 1997–2005, was 83% overall, from 75% in Latin American countries to 92% in North America. Slightly lower susceptibilities (77%) were seen in North American isolates from patients in the hematology-oncology services. Of all isolates tested, 100% were susceptible to all of the echinocandins, but there were some decreased susceptibilities seen to amphotericin and 5-FC [21]. Isolates are known to be intrinsically resistant to the azoles, although in vitro data may not give MICs in the resistant range. In one set of ARTEMIS global trending data, the percent resistance had increased in vitro from 66% in 2000 to 79% in 2005 [20]. Likewise in the SENTRY data, with 29 isolates of *C. krusei* tested, 3.5% of isolates were susceptible, 79.3 fell in the S-DD range, and 17.2% in the resistant range [17]. In these same SENTRY data, 100% of the 29 isolates were found susceptible to anidulafungin and 97% to caspofungin; 93% were susceptible to amphotericin and 93% susceptible to voriconazole. Only 3.4% were susceptible to 5-FC, which is a common characteristic of most studies reporting on *C. krusei* susceptibility [17]. For reporting purposes, clinical laboratories should report results versus fluconazole as resistant for *C. krusei* regardless of the in vitro MIC results.

5.1.6   *C. lusitaniae*

Candida lusitaniae is often considered to be non-susceptible to amphotericin because of a few reports of cases of meningitis and fungemia in which the MICs of the isolates were very high [12, 16]. The susceptibility tests done were agar-dilution based and without any CLSI/NCCLS guidelines to follow. In the SENTRY data, using CLSI guidelines for testing and interpretation, 14 isolates of *C. lusitaniae* were found 100% susceptible to amphotericin-B, the echinocandins, and 5-FC. In addition, 93% were fully susceptible to fluconazole and 100% to voriconazole. Only 64%, however, were susceptible to itraconazole [17]. A review article of other published susceptibility data reported that in 70 isolates of *C. lusitaniae*, the $MIC_{90\%}$ was <0.5–2, indicating that there were a few amphotericin-B-resistant strains.

In that same review, 114 (100%) isolates were susceptible to caspofungin, and 107/107 were susceptible to fluconazole and 27/27 susceptible to voriconazole [8]. Looking at bloodstream isolates, Ostrosky-Zeichner found no resistance to amphotericin in 20 *C. lusitaniae* tested in the USA [18]. Lastly, there is a paper describing the colonial morphology changes in an isolate of *C. lusitaniae* in a patient with systemic infection and along with these morphologic changes, the emergence of amphotericin-B resistance [15]. In this day of newer antifungal agents, there are other choices than amphotericin-B for treatment of potentially resistant yeasts or those that might become resistant. Alternatively, if amphotericin-B is being considered, a susceptibility test can be done on that particular isolate [13].

5.1.7   *C. guilliermondii*

C. guilliermondii is rarely isolated in clinical laboratories, except in Latin America where it has been reported as the sixth most common *Candida* spp. The ARTEMIS Surveillance data has reported on the response to azoles and echinocandins of >1,000 clinical isolates of *C. guilliermondii* from around the world between 2001 and 2003. Only 71% of the isolates were susceptible to fluconazole as compared to 91% susceptibility to voriconazole. Most of the isolates (96% of those tested) were susceptible to caspofungin [25]. In a review study of many published articles, the $MIC_{90\%}$ of 58 isolates of *C. guilliermondii* was 0.25–0.5 µg/ml [8].

5.1.8   *C. rugosa*

Candida rugosa is seen in some countries outside of the USA, including Canada and Latin American countries with higher frequencies. It has been shown to have decreased susceptibility to azoles, especially fluconazole, and should be considered as another potential fluconazole-resistant species like *C. glabrata* and *C. krusei* [19].

5.2   *Cryptococcus* spp.

5.2.1   *C. neoformans*

Cryptococcus neoformans is susceptible to amphotericin, 5-FC, and the azoles; however, all strains will be intrinsically resistant to the echinocandins. Rarely strains of *C. neoformans* may be initially resistant to the azoles, or 5-FC or develop that resistance during therapy. A study of global trends in the susceptibility of *C. neoformans* from 1990 to 2004 in five geographic regions of the world reported that resistance to amphotericin-B, 5-FC, and fluconazole was ≤1%. However, only 75% of North American strains were fully susceptible to fluconazole (MIC ≤8 μg/ml) compared to 94–100% in the other four regions (Africa, Latin America, Europe, and Asia-Pacific). Susceptibility to voriconazole and posaconazole was good and similar throughout all regions. Susceptibility to 5-FC increased from 34% in 1990–1994 to 66% in 2000–2004; susceptibility to fluconazole also increased over the 15 years surveyed [26]. In another in vitro study, isolates from the USA, Thailand, and Malawi demonstrated no evidence of resistance to amphotericin-B, 5-FC, fluconazole, and itraconazole in a 2004 report [1]. In a surveillance study of drug susceptibility testing of almost 500 C. neoformans isolates in South Africa for two time periods, 2002–2003 and 2007–2008, only three (0.6%) of the isolates had an MIC to fluconazole ≥16 μg/ml and only in the first time period studied. All isolates were susceptible to amphotericin-B and had low MICs to voriconazole and posaconazole. When isolates from the same patient were collected during therapy, only one case was detected in which the MIC to fluconazole was significantly higher than at the time of initial isolation in the patient [11]. In another study of fluconazole and amphotericin-B susceptibility testing of sequential strains during treatment of the patient, all isolates continued to have very low MICs to amphotericin (≤1 mcg/ml); in five cases of relapse with continued positive cultures, one isolate became resistant (MIC > 64 mcg/ml) and four others had MICs in the S-DD range. There appeared to be no adverse clinical outcomes as a result of this, however [2]. The $MIC_{90\%}$ of 100 clinical isolates of C. neoformans from Taiwan were reported to have MICs indicating susceptibility to fluconazole, itraconazole, 5-FC, voriconazole, and amphotericin-B. There was only one identified C. gattii in this collection [14].

5.2.2   C. gattii

C. gattii isolates have recently become emergent pathogens throughout the world. In a study from India, in which 308 clinical and environmental isolates of C. neoformans var. grubii and C. gattii serotype B were compared for results in susceptibility testing with a standard microbroth dilution method, only two isolates of C. neoformans var. grubii were found resistant to 5-FC. C. gattii strains were found to be significantly less susceptible to fluconazole, itraconazole, and voriconazole, but both responded similarly as susceptible to amphotericin-B and 5-FC. In addition, the environmental isolates of C. neoformans var. grubii were found to be significantly less susceptible to fluconazole, itraconazole, and 5-FC as compared to clinical strains [6]. In a Brazilian study, 80 clinical isolates of C. neoformans var. grubii and 4 isolates of C. gattii were all found susceptible to amphotericin-B, fluconazole, itraconazole, and voriconazole [28]. Twenty-three clinical isolates of C. gattii were compared to C. neoformans strains in Spain and were found to be more susceptible to amphotericin-B and 5-FC; however, fluconazole and other azoles demonstrated higher MICs for C. gattii compared to *C. neoformans* [10]. Another study in Colombia comparing the two serotypes of *C. neoformans* found that there were similarities in interpretive criteria; however, the MICs were actually higher with *C. gattii* strains [7].

A study of susceptibilities in Nairobi, Kenya, points out the necessity of developing criteria in one's geographic area for antifungal susceptibility testing patterns. In 80 clinical isolates of *C. neoformans* (75/80 = var. *grubii*, 3 var. *neoformans* and 2 var. *gattii*). Hundred percent of all strains were susceptible to amphotericin, 21% resistant to 5-FC, 24% susceptible and 65% S-DD to fluconazole, 11% resistant to fluconazole (using breakpoints for *Candida* sp). Increased prophylactic use of the azoles in HIV⁺ individuals may have increased the overall resistance seen in these isolates [4].

5.2.3   Other *Cryptococcus* spp.

There are not many studies that report on the in vitro susceptibility testing results with species of *Cryptococcus* spp. other than *C. neoformans*. In a study out of Spain in 2010, 122 isolates were reported on (24 such isolates from their institution and 98 other strains from a review of the literature). Included were *C. albidus* (46%), *C. laurentii* (32%), and the other 22% composed of *C. uniguttulatus, C. humicola, C. curvatus*, and *C. luteolus*. Most remained susceptible to amphotericin-B, but non-susceptible to the echinocandins and 5-FC. Fluconazole was not susceptible against many strains; in particular, it was inactive versus *C. albidus, C. uniguttulatus*, and *C. laurentii*. Resistance was seen for some species versus voriconazole, itraconazole, and posaconazole [3].

5.3   Other Yeasts

5.3.1   *Rhodotorula* spp.

Rhodotorula spp. are not commonly isolated from clinical specimens, but there is some available data about antimicrobial susceptibility testing of isolates. There is an intrinsic resistance to both the azoles and echinocandins. Results of 29 isolates from one laboratory in Spain were combined with a literature review of susceptibility results of an additional 102 clinical isolates of *Rhodotorula* spp. including *R. mucilaginosa, R. glutinis,* and other *Rhodotorula* spp. An MIC of ≤1 μg/ml was demonstrated versus amphotericin-B for all isolates. Good activity was seen with 5-FC as well. Most isolates had very high MICs to fluconazole; although some isolates did have lower MICs to itraconazole and voriconazole, the authors concluded that the activity of these azoles was poor [9]. Similarly, the ARTEMIS Global Surveillance Study reported on results from >380 isolates of *Rhodotorula* spp. from five geographic locations during years 1997–2007. Fifty to more than eighty percent of isolates were resistant to fluconazole using a disk diffusion CLSI method; 40–69% were also resistant to voriconazole. The least amount of resistance was reported in isolates from Asia-Pacific regions, as low as 15% versus voriconazole, although the authors still concluded that azoles should be considered resistant against *Rhodotorula* spp. [24].

5.3.2   Trichosporon spp.

Most isolates of *Trichosporon asahii* are found to be resistant to amphotericin-B, with MICs ≥2 μg/ml and often much higher than that. Susceptibility patterns of a number of *Trichosporon* spp. were reported from Chile in 2005, using conventional methods as well as molecular methods for detection of resistance. The majority of 15 isolates of *T. asahii* had high MICs to amphotericin-B as compared to predominantly susceptible MICs for *T. inkin* and *T. mucoides*, two other commonly isolated species of *Trichosporon*. In that same study, fluconazole performed poorly against *T. asahii*, but voriconazole showed good activity [27]. A Brazilian study also demonstrated that nearly 50% of the *T. asahii* strains would be considered resistant (MICs ≥2 μg/ml), but there was more activity by all azoles, although voriconazole did again perform best. All 22 strains of *Trichosporon* spp. had MICs ≥2 μg/ml versus the echinocandin, caspofungin, and also high MICs to 5-FC [5]. Across the five geographic regions of the ARTEMIS Global Surveillance Study, 0–12% resistance to fluconazole was seen by ~1,000 *Trichosporon* spp. and that included *T. asahii*. The percent resistance to voriconazole was very low throughout all regions [24].

5.3.3   Saccharomyces spp.

Saccharomyces cerevisiae and other species of *Saccharomyces* rarely cause human infections; however, they have been reported in cases of fungemia and even the rare case of endocarditis. In the ARTEMIS Global Surveillance Study, <6% of 470 isolates of *S. cerevisiae* tested between 2005 and 2007 were resistant to fluconazole and <3% to voriconazole. When the data was stratified by regions, up to 24% of the Latin American and Asia-Pacific isolates were found to be fluconazole resistant and >10% voriconazole resistant, so susceptibility testing would be important in those regions if azoles were being considered for treatment [24]. In a review of results of 48 S. *cerevisiae* isolates, the $MIC_{90\%}$ to amphotericin-B was 1.0 μg/ml and to 5-FC, 0.12–0.25; the $MIC_{90\%}$ of 46 strains versus anidulafungin and 10 versus caspofungin was ≤2 μg/ml, which should be considered susceptible for all of these antifungal agents [8].

Table 5.1 Common in vitro susceptibility patterns for yeasts

	Ampho-B	5-FC	Flucon	Voricon	Posacon	Caspo	Micafun	Anidulafun	Should isolate be tested
C. albicans	S	S	S	S	S	S	S	S	Routinely not; only upon request
C. glabrata	S	S	NS	S-usually	S	S	S	S	Yes
C. parapsilosis	S[a]	S	S	S	S	S[a]	S[a]	S[a]	With failures
C. tropicalis	S[a]	S[b]	S	S	S	S	S	S	With failures
C. krusei	S	R	R	S	S	S	S	S	No; report as Fluconazole R
C. lusitaniae	S[c]	S	S	S	S	S	S	S	Not usually
C. guilliermondii	S	S	NS	S	NA	S	NA	NA	Yes
C. rugosa	NA		NS	NA	NA	NA	NA	NA	Yes
Crypto. neoformans	S	S	S	S	S	R	R	R	Not usually
Other Cryptococcus	S	S	May be NS	May be NS	May be NS	R	R	R	Yes, if clinically significant
Rhodotorula	S	S	NS	NS	NS	NS	NS	NS	Yes, if clinically significant
Trichosporon	NS	NS	NS	S-usually	NA	NS	NS	NS	Yes, if clinically significant
Saccharomyces spp.	S	S	Usually NS	S	NA	S	NA	S	Yes, if clinically significant

S susceptible, NS non-susceptible, R resistant, NA not available

[a] Rare resistance reported in some studies

[b] Often has higher MICs than C. albicans or C. parapsilosis

[c] Not usually non-susceptible initially, but isolates may develop decreased susceptibility with treatment

References

1. Archibald LK, Tuohy MJ, Wilson DA et al (2004) Antifungal susceptibilities of *Cryptococcus neoformans*. Emerg Infect Dis 10:143–145
2. Arechavala A, Ochiuzzi ME, Borgina MD et al (2009) Fluconazole and amphotericin B susceptibility testing of *Cryptococcus neoformans*: results of a minimal inhibitory concentrations against 265 isolates from HIV-positive patients before and after two or more months of antifungal therapy. Rev Iberoam Micol 26:194–197
3. Bernal-Martinez L, Gomez-Lopez A, Castelli MV et al (2010) Susceptibility profile of clinical isolates of non-*Cryptococcus neoformans*/non-*Cryptococcus gattii Cryptococcus* species and literature review. Med Mycol 48:90–96
4. Bii CC, Makimura K, Abe S et al (2007) Antifungal susceptibility of *Cryptococcus neoformans* from clinical sources in Nairobi, Kenya. Mycoses 50:25–30
5. Chagas-Neto TC, Chaves GM, Melo ASA, 47 et al (2009) Bloodstream infections due to *Trichosporon* spp.: species distribution, *Trichosporon asahii* genotypes determined on the basis of ribosomal DNA intergenic spacer 1 sequencing, and antifungal susceptibility testing. J Clin Microbiol 47:1074–1081
6. Chowdhary A, Randhawa HS, Sundar G et al (2011) In vitro antifungal susceptibility profiles and genotypes of 308 clinical and environmental isolates of *Cryptococcus neoformans* var *grubii* and *Cryptococcus gattii* serotype B from north-western India. J Med Microbiol 60:961–967
7. DeBedout C, Ordonez N, Gomez BL et al (1999) In vitro antifungal susceptibility of clinical isolates of Cryptococcus neoformans var neoformans and C. neoformans var. gattii. Rev Iberoam Micol 16:36–39
8. Espinel-Ingroff A (2003) In vitro antifungal activities of anidulafungin and micafungin, licensed agents and the investigational triazole posaconazole as determined by NCCLS methods for 12,052 fungal isolates: review of the literature. Rev Iberoam Micro 20:121–36
9. Gomez-Lopez A, Mellado E, Rodriquez-Tudela JL et al (2005) Susceptibility profile of 29 clinical isolates of *Rhodotorula* spp. and literature review. J Antimicrob Chemother 55:312–6
10. Gomez-Lopez A, Zaragoza O, Dos Anjos Martins M, Melhern MC et al (2008) In vitro susceptibility of Cryptococcus gattii clinical isolates. Clin Microbiol Infect 14:727–730
11. Govender NP, Patel JVM, Chiller TN et al (2011) Trends in antifungal drug susceptibility of *Cryptococcus neoformans* obtained through population-based surveillance, South Africa, 2002–2003 and 2007–2008. Antimicrob Agents Chemother 55:2606–2611, Mar 28. epub ahead of print
12. Guinet R, Chanas J, Goullier A et al (1983) Fatal septicemia due to amphotericin B-resistant *Candida lusitaniae*. J Clin Microbiol 18:443–444
13. Hawkins JL, Baddour LM (2003) *Candida lusitaniae* infections in the era of fluconazole availability. Clin Infect Dis 36(3):14–18
14. Liaw SJ, Wu HC, Hsueh PR (2010) Microbiological characteristics of clinical isolates of *Cryptococcus neoformans* in Taiwan: serotypes, mating types, molecular types, virulence factors, and antifungal susceptibility. Clin Microbiol Infect 16:696–703
15. McClenny NB, Fei H, Baron EJ et al (2002) Change in colony morphology of *Candida lusitaniae* in association with development of amphotericin B resistance. Antimicrob Agents Chemother 46:1325–1328
16. Merz WG (1984) *Candida lusitaniae*: frequency of recovery, colonization, infection, and amphotericin B resistance. J Clin Microbiol 20:1194–1195
17. Messer A, Moet GJ, Kirby JT et al (2009) Activity of contemporary antifungal agents, including the novel echinocandin, anidulafungin, tested against Candida spp., Cryptococcus spp., and Aspergillus spp.: report from the SENTRY Antimicrobila Surveillance Program (2006 to 2007). J Clin Microbiol 47:1942–1946

18. Ostrosky-Zeichner L, Rex JH, Pappas PG et al (2003) Antifungal susceptibility survey of 2000 bloodstream *Candida* isolates in the United States. Antimicrob Agents Chemother 47: 3149–3154
19. Pfaller MA, Diekema DJ, Colombo AL et al (2006) *Candida rugosa*, an emerging fungal pathogen with resistance to azoles: geographic and temporal trends from the ARTEMIS DISK antifungal surveillance program. J Clin Microbiol 44:3578–3582
20. Pfaller MA, Diekema DJ, Gibbs DL et al (2007) Results from the ARTEMIS Disk Global Antifungal Surveillance Study, 1997 to 2005: 8.5 year analysis of susceptibilities of Candida species and other yeast species to fluconazole and voriconazole determined by CLSI standardized disk diffusion testing. J Clin Microbiol 45:1735–1745
21. Pfaller MA, Diekema DJ, Gibbs DL (2008) *Candida krusei*, a multidrug-resistant opportunistic fungal pathogen: geographic and temporal trends from the ARTEMIS DISK Antifungal Surveillance Program, 2001 to 2005. J Clin Microbiol 46:515–521
22. Pfaller MA, Diekema DJ, Gibbs DL et al (2008) Geographic and temporal trends in isolation and antifungal susceptibility of *Candida parapsilosis*: a global assessment from the ARTEMIS DISK Antifungal Surveillance Program, 2001 to 2005. J Clin Microbiol 86:842–849
23. Pfaller MA, Diekema DJ, Gibbs DL et al (2010) Geographic variation in the frequency of isolation and fluconazole and voriconazole susceptibilities of *Candida glabrata*: an assessment from the ARTEMIS DISK Global Antifungal Surveillance Program. Diagn Microbiol Infect Dis 672:162–171
24. Pfaller MA, Diekema DJ, Gibbs DL et al (2009) Results from the ARTEMIS DISK global surveillance study, 1997–2007: 10.4 year analysis of susceptibilities of noncandidal yeast species to fluconazole and voriconazole determined by CLSI standardized disk diffusion testing. J Clin Microbiol 47:117–123
25. Pfaller MA, Diekema DJ, Mendez M et al (2006) *Candida guilliermondii*, an opportunistic fungal pathogen with decreased susceptibility to fluconazole: geographic and temporal trends from the ARTEMIS DISK antifungal surveillance program. J Clin Microbiol 44:3551–3556
26. Pfaller MA, Messer SA, Boyken L et al (2005) Global trends in the antifungal susceptibility of *Cryptococcus neoformans* (1990–2004). J Clin Microbiol 43:2163–2167
27. Rodriguez-Tudela JL, Diaz-Guerra TM, Mellado E et al (2005) Susceptibility patterns and molecular identification of *Trichosporon* species. Antimicrob Agents Chemother 49:4026–4034
28. Souza LK, Souza AH, Costa CR et al (2010) Molecular typing and antifungal susceptibility of clinical and environmental *Cryptococcus neoformans* species complex isolates in Goiania, Brazil. Mycoses 53:62–67

Chapter 6
Usual Susceptibility Patterns of Common Moulds and Systemic Fungi

Gerri S. Hall

Abstract The in vitro susceptibility patterns of some of the common moulds such as *Aspergillus* spp., *Fusarium* spp., and other hyaline moulds; Zygomycetes, and dematiaceous moulds such as *Alternaria* spp., *Curvularia* spp., *Exophiala* spp., and other black moulds are described in this chapter. For some isolates such as *A. fumigatus* and *A. flavus,* results are fairly predictable, and isolates are often susceptible to all agents tested; others such as *Fusarium* sp. may demonstrate less predictable patterns that require that susceptibility be performed for each isolate of significance. In this chapter, we describe the most common susceptibility patterns as found in the published literature, including some geographic differences throughout various parts of the world where results have been reported.

6.1 Hyaline Moulds

6.1.1 Aspergillus spp.

6.1.1.1 Aspergillus fumigatus

Aspergillus fumigatus is usually susceptible in vitro to most antifungal agents. In the SENTRY antimicrobial surveillance 2006–2007, 29 isolates of *A. fumigatus* were reported to be 100% susceptible to itraconazole, voriconazole, posaconazole, caspofungin, and anidulafungin; the $MIC_{90\%}$ of the 29 isolates vs. amphotericin B was 2 μg/ml, and only 71.4% were noted to be susceptible to amphotericin with an MIC of ≤1 μg/ml [27]. Data from the same surveillance group in 2009 reported on 40

G.S. Hall, Ph.D. (✉)
Section of Clinical Microbiology, Department of Clinical Pathology,
Cleveland Clinic, Cleveland, OH 44195, USA
e-mail: hallg@ccf.org

G.S. Hall (ed.), *Interactions of Yeasts, Moulds, and Antifungal Agents:*
How to Detect Resistance, DOI 10.1007/978-1-59745-134-5_6,
© Springer Science+Business Media, LLC 2012

A. fumigatus isolates, and again, 100% were susceptible the echinocandins, caspofungin, micafungin, anidulafungin, and the azoles itraconazole, voriconazole, and posaconazole. No information was given on amphotericin testing in this latter series [32]. A study of 375 isolates in Spain in 2006 reported an $MIC_{90\%}$ to amphotericin of 0.5 µg/ml, which is usually an indication of susceptibility [12]. There is some reported variability of in vitro amphotericin susceptibility among strains of *A. fumigatus*.

In a UK study of data collected in 2008 and 2009, of 230 isolates of A. *fumigatus*, 64 (28%) were azole resistant, 14% in 2008, and 20% of patients in 2009 had resistant isolates, respectively. During this period, 62 of 64 (97%) were itraconazole resistant, 2 of 64 (3%) only were voriconazole resistant, and 78% of cases were multiazole resistant [7]. In a study by Pfaller et al. in which triazole cross-resistance was being specifically determined, 553 isolates of *A. fumigatus* had susceptibilities performed following CLSI M38-A standard and interpretations were designated as ≤1 µg/ml = S and ≥4 µg/ml = R. For itraconazole, 93% of the *A. fumigatus* were susceptible, and ≥99% were susceptible to posaconazole, voriconazole, and ravuconazole [33]. There have been a number of studies which suggest that although azole resistance in *A. fumigatus* is rare, it is increasing; using itraconazole resistance as a marker of this resistance seems warranted since no resistance in voriconazole, posaconazole, or ravuconazole has been seen in itraconazole susceptible strains [30]. Cross-resistance to azoles was studied in Spain among 393 isolates of *A. fumigatus*, including 32 itraconazole-resistant strains. These authors showed that specific mutations in the cyp51A locus could result in total cross-resistance among all azoles, and like others suggested that determination of itraconazole resistance was important as a marker of potential azole resistance development [35].

6.1.1.2   *A. flavus* and *A. niger*

Aspergillus flavus is often the second most common clinically significant *Aspergillus* spp. isolated in clinical laboratories. It usually demonstrates a susceptibility pattern similar to *A. fumigatus*, i.e., susceptible to amphotericin, itraconazole, voriconazole, and the echinocandins [12, 16, 31, 33]. Only 70% of 30 isolates of *A. flavus* reported by Diekema et al., from an Iowa study in 2003, were susceptible to amphotericin, using <1 µg/ml a s breakpoint for susceptibility; in addition, >96.7% were susceptible to the azoles (itraconazole, voriconazole, and posaconazole) and 100% to caspofungin, using a susceptibility breakpoint of <1 µg/ml MIC or MEC respectively [14]. *Aspergillus niger*, the third most frequent isolate in clinical specimens, has been reported to be susceptible to amphotericin ($MIC_{90\%}$ of 0.25 µg/ml for 55 strains tested in a Spanish study and 100% susceptible in a US study in 2003) [12, 14]. *Aspergillus niger* has been shown to have higher MICs to the some of the triazoles as compared to *A. flavus* and *A. fumigatus*. In the 2003 US study, 45% of 29 A. niger isolates only were susceptible to itraconazole, however, and 66% to voriconazole as compared to 100% susceptibility to posaconazole, using <1 µg/ml as breakpoint for all. All 29 isolates were susceptible to caspofungin [14, 18]. In the Pfaller study examining cross-resistance of *Aspergillus* spp. to the azoles, only 41%

of 59 *A. niger* strains had an MIC ≤ 1 μg/ml to itraconazole compared to the >93% for *A. fumigatus* and *A. flavus* [33]; likewise in the Cuenca-Estrella study, the MIC$_{90\%}$ of 55 *A. niger* isolates was 4.0 μg/ml, which would be considered an indication of resistance using the CLSI interpretations [12]. In the same Pfaller study of cross-resistance, for voriconazole, 97% of the isolates were susceptible (MIC ≤ 1 μg/ml), similar to that of *A. flavus* and *A. fumigatus*; and 92% for posaconazole [33]. Eighty-three strains of *A. niger* had an MIC$_{90\%}$ < 0.25 μg/ml for all three echinocandins; these same isolates of *A. niger* had MIC$_{90\%}$ of >8, 2.0, and 0.5 μg/ml, respectively vs. itraconazole, voriconazole, and posaconazole, demonstrating some variability vs. the azoles for *A. niger* [13].

6.1.1.3   *A. versicolor*

A. versicolor is a less frequently isolated species of Aspergillus. Twenty strains were tested in an Iowa study in 2003, and using <1 μg/ml (MIC for amphotericin and azoles and MEC for echinocandins) as a breakpoint for susceptibility, 80% were susceptible to amphotericin, 91% to caspofungin, only 65% to itraconazole, but greater than 90% for voriconazole and posaconazole [14]. In the Cuenca-Estrella study in Spain, with 13 strains of *A. versicolor*, the MIC$_{90\%}$ for itraconazole was ≤1.0 μg/ml for itraconazole and posaconazole and was 2.0 μg/ml for voriconazole. The MIC$_{90\%}$ for amphotericin was <2.0 μg/ml, suggesting that some isolates were potentially nonsusceptible [12]. In a study the following year with 12 strains of *A. versicolor*, results were the same vs. amphotericin and the azoles; in addition, the MIC$_{90\%}$ for caspofungin was 2.0 μg/ml, and <0.06 μg/ml for micafungin and anidulafungin [13].

6.1.1.4   *A. terreus*

A. terreus strains have been shown to be more resistant to amphotericin than other *Aspergillus* spp. Seventy-four isolates in Spain were tested, and MIC$_{90\%}$ to amphotericin was 8.0 μg/ml, considerably above the nonsusceptible breakpoint which is usually considered ≥2.0 μg/ml. In the same study, the MIC$_{90\%}$ to the azoles, itraconazole, voriconazole, and posaconazole were all susceptible at ≤1.0 μg/ml. In a US study from Iowa in 2003, using <1.0 μg/ml as a susceptible breakpoint for all agents, 100% of 16 strains of *A. terreus* were susceptible to caspofungin, and the azoles, itraconazole, voriconazole, and posaconazole; however, over 37% were susceptible to amphotericin [14]. A study from Spain, using EUCAST testing methods, with 48 clinical isolates of *A. terreus*, obtained a range of MICs vs. amphotericin of 0.5–8 μg/ml, with few isolates in which the MIC was <1 μg/ml. The MIC$_{90\%}$ was 2.0 μg/ml, which was lower than found in studies employing CLSI testing methods. MICs to voriconazole were high, with an MIC$_{90\%}$ of 2.0 μg/ml; for posaconazole, the MIC$_{90\%}$ was 0.12 μg/ml; for itraconazole, 0.5 μg/ml; and for terbinafine, 0.5 μg/ml. The authors concluded that there was a difference between CLSI and EUCAT methods especially with voriconazole; however, without clinical outcomes correlations, it would be impossible to know which was the correct MIC [18, 24].

6.1.1.5 *A. nidulans*

For *A. nidulans*, in vitro data from 29 strains from Spain in 2006 showed that all were resistant (>8.0 µg/ml) to itraconazole, voriconazole, and posaconazole, and had an $MIC_{90\%}$ to amphotericin of 2.0 µg/ml [12]. Interestingly, in 2009, 20 isolates of *A. nidulans* were all found susceptible (<1 µg/ml) to the azoles and again a higher MIC to amphotericin with $MIC_{90\%}$ at 4.0 µg/ml [13]. In USA in 2008, Espinel-Ingroff reported that for 13 isolates of *A. nidulans*, most were susceptible to the itraconazole and voriconazole, with an $MIC_{90\%}$ to amphotericin of 2.0 µg/ml [16].

6.1.1.6 Epidemiological Cutoff Values for *Aspergillus* spp.

For the *Aspergillus* spp., another way of examining susceptibility to the azoles that has been used is to determine the epidemiological cutoff values (ECVs) to distinguish wild-type strains from those that harbor resistance mutations [34]. Use of these ECVs were applied to over 1,700 strains of *Aspergillus* spp. from 63 different worldwide centers from 2001 to 2009 in order to determine the frequency of non-wild type strains. For all of the wild-type strains of *A. fumigatus, A. flavus, A. terreus, A. niger, A. versicolor*, and *A. nidulans*, the percentage of isolates at or below the epidemiological cutoff was ≥97.8%. For the non-wild type strains, the percentage of MIC above ECV ranged from 0% to 17.3% for itraconazole, 0–3.2% for voriconazole, and 0–5.1% for posaconazole. *A. niger* had the highest percentage of MICs>ECV for itraconazole [32, 34]. Ninety-nine percent of >2,500 isolates of *A. fumigatus* had MICs less than the epidemiological cutoff value (ECV) for itraconazole, and 99% and 98% respectively for posaconazole (>1,600 strains) and voriconazole (>2,700 strains). Isolates in this analysis were obtained from US, Spain, and the UK. For hundreds of isolates of *A. flavus, A. terreus, A. niger, A. nidulans* and *A. versicolor*, results were similar, with percentage less than ECV > 95% for the three azoles. There was a recommendation that if an isolate of one of these species of *Aspergillus* spp. had an MIC to posaconazole above 0.25 µg/ml, reduced susceptibility to this triazole should be considered, and standard dosing regimens may have to be increased [17].

6.1.1.7 Synergy and *Aspergillus* spp.

A study looking at the possible synergism of combining voriconazole and micafungin against *Aspergillus* and other moulds showed a 79% synergism effect with *A. fumigatus* at an $MIC_{50\%}$ level. Forty percent (10/24) of the *A. flavus* and 2/6 isolates of *A. niger* and *A. nidulans* also demonstrated a synergistic effect. None of the *Aspergillus* spp. tested demonstrated any antagonism when voriconazole and micafungin were combined [22].

6.1.2   *Fusarium spp.*

There is great variability among the species of *Fusarium* in regard to their in vitro susceptibility to antifungal agents. In general, all *Fusarium* are more resistant than are the *Aspergillus* spp. Data from a study in Spain with 44 strains of *Fusarium* spp. showed that the three species (*F. solani, F. oxysporum,* and *F. verticillioides*) were resistant to the azoles, itraconazole, voriconazole, and posaconazole, with $MIC_{90\%} \geq$ 8 µg/ml. *F. solani*, and *F. verticillioides* had a high $MIC_{90\%}$ to amphotericin ($\geq$4 µg/ ml), whereas the $MIC_{90\%}$ for *F. oxysporum* was 1.0 µg/ml, indicating susceptibility to amphotericin [12]. Another Spanish study in 2009 from the same lab reported very high MICs for the three above species of *Fusarium* as well as 19 *F. prolifera-tum* sp. [5]. Eleven *Fusarium* spp. from USA and Canada reported in a study from Iowa in 2003, using percentage of isolates <1 µg/ml to indicate susceptibility, demonstrated that 82% were susceptible to amphotericin, and 18% to voriconazole and posaconazole; no isolates were considered susceptible to itraconazole and caspofungin [14]. All of the 57 isolates of *Fusarium* spp. reported from Mexico in 2005 had an $MIC_{90\%} > 1$ µg/ml for itraconazole, voriconazole, and posaconazole, an indication that the majority were resistant to the azoles in vitro [21].

In one study of ocular isolates of *Fusarium* spp., *F. solani* demonstrated consistently higher MICs to itraconazole, voriconazole, and posaconazole than did species of *F. oxysporum*. Most *Fusarium* spp. had high MICs to amphotericin, natamycin, and the echinocandins in that same study [23]. Likewise, when in vitro pharmacodynamics was compared to noncultural method endpoints, *F. solani* was found cross-resistant to amphotericin B, itraconazole, and voriconazole [25]. A study in Spain in 2008 used molecular methods to identify species and then compare the identification to their antifungal susceptibility patterns (Table 6.1). The most frequent isolates were *F. solani, F. oxysporum, F. proliferatum,* and *F. verticillioides*. Amphotericin B was the only drug with low MICs for all species; the azoles (itraconazole, voriconazole, and posaconazole) all demonstrated high MICs as did terbinafine [1]. In a Spanish study using methods approved by AFST-EUCAST, the results vs. amphotericin of *F. solani* (32 strains) and *F. oxysporum* (19 strains) were an $MIC_{90\%}$ of 2.0 µg/ml; for 19 strains of *F. proliferatum* and 11 of *F. verticillioides*, the $MIC_{90\%}$ was 4.0 µg/ml, both of which would be considered, according to CLSI 38-A, as intermediate. All three azoles (itraconazole, voriconazole, and posaconazole) demonstrated $MIC_{90\%}$ of >4 µg/ml (interpreted as resistant); poor activity was reported vs. the three echinocandins, with $MIC_{50\%}$ and $MIC_{90\%}$ >16 µg/ml [9, 13].

Susceptibility testing is needed for each isolate when clinically relevant since results differ between species and are also dependent on the method for identification and potentially the country of origin of the isolate. Resistance to most antifungal agents will be anticipated, however, for many of the *Fusarium* spp.

Comparison of CLSI methods for antifungal susceptibility testing was compared to the XXT reduction assay in conjunction with fluorescent morbidity staining for two strains each of *Fusarium solani* and *F. oxysporum*. One of the *F. solani* isolates had an MIC that was 1 µg/ml to amphotericin; the other four strains had MICs > 1 µg/ml and MFCs > 2 μg/ml. One of the *F. solani* had MICs and MFCs to amphotericin that were very high (>8 μg/ml). The four *Fusarium* spp. had high MICs to itraconazole (≥8 μg/ml) and voriconazole (≥2 μg/ml). The XXT assay was used to determine hyphal damage to characterize the pharmacodynamics of these three antifungal agents. By using these assays, itraconazole was seen to have less activity than amphotericin B or voriconazole against the four *Fusarium* spp. Amphotericin B showed good activity against ¾ *Fusarium* spp. even though fairly high failure rates usually exist when attempting to treat *Fusarium* infections with amphotericin; data from the in vitro MIC tests indicated nonsusceptibility for three of the four isolates. Amphotericin exhibits a nonlinear concentration-dependent binding to proteins in serum and tissue, and this increases with increasing drug concentrations. Voriconazole demonstrated good activity against the four *Fusarium* strains [25].

6.1.3   Scedosporium spp.

Scedosporium apiospermum and *S. prolificans* are intrinsically resistant to amphotericin and variably resistant to the azoles. In a study in Spain, 65 isolates had a predictably high $MIC_{90\%}$ of >16 μg/ml to amphotericin, an $MIC_{90\%}$ ≥ 8 μg/ml for itraconazole and posaconazole, and $MIC_{90\%}$ of 4.0 μg/ml for voriconazole. Thirty-seven isolates of *S. prolificans* had $MIC_{90\%}$ > 32 μg/ml and >8 μg/ml for amphotericin and azoles (itraconazole, voriconazole, and posaconazole), respectively [10]. Using the Sensititre YeastOne panel for performing MICs to voriconazole, Linares, also in a study in Spain, with six isolates of *S. apiospermum*, reported an $MIC_{90\%}$ of 0.5 μg/ml, which would be considered susceptible [26]. The echinocandins were tested against 36 *S. apiospermum* and 17 isolates of *S. prolificans* with the following results of the $MIC_{90\%}$ for each: caspofungin and micafungin were >16 μg/ml for both species, and anidulafungin had an $MIC_{90\%}$ of 4.0 μg/ml vs. *S. apiospermum*, and >16 μg/ml for *S. prolificans*. The only suggested breakpoint for susceptibility for the echinocandins listed in the CLSI document M38-A is for caspofungin (≥4 μg/ml=*R*), but if that carries over to any echinocandin, most *Scedosporium* spp. would be considered resistant to the echinocandins as a class along with resistance to the azoles and amphotericin [13].

    Activities of 35 combinations of antifungal agents against *Scedosporium* spp. were analyzed by a checkerboard microdilution design and the summation of fractional concentration index. An average indifferent effect was detected apart from combinations of azole agents and echinocandins against *Scedosporium apiospermum*. Antagonism was absent for all antifungal combinations against both species [10]. The in vitro interaction between amphotericin B and micafungin against 36 isolates of *Scedosporium* spp. has been evaluated using checkerboard assays and the minimal effective concentration endpoint. Synergy was found for 82.4% of *Scedosporium prolificans* isolates and for 31.6% of *Scedosporium apiospermum* isolates. Antagonism was not observed [38].

The antifungal susceptibilities of the *Pseudallescheria boydii* complex were reported from a Spanish laboratory in 2006. Eighty-four isolates of eight species in this complex were tested against 11 antifungal agents. The species were as follows: 30 strains of P. boydii, 4 each of *P. minutispora* and *P. angusta*, 6 *P. ellipsoidea*, 2 *P. fusoidea*, 7 *Scedosporium aurantiacum*, 26 of what is a cryptic species referred to as Clade 4, and 5 of Clade 3. Amphotericin was not active against any of the isolates. Voriconazole was the most active drug, with $MIC_{90\%} \leq 1$ µg/ml for all except S. *aurantiacum*. Results with posaconazole were similar, and again, S. *aurantiacum* was not susceptible. All 84 isolates were resistant to micafungin. The authors concluded that voriconazole seemed to be the most effective agent; however, they suggested that proper identification of the species within the complex should be done in order to provide correct information for treatment, especially if S. *aurantiacum* was the identity of the mould [20]. PCR and DNA sequencing of the internal transcribed spacer (ITS) region of 46 clinical isolates of morphologically identified *P. boydii* and S. *apiospermum* was performed along with antifungal susceptibility testing according to CLSI recommendations. Four of the 46 isolates were identified molecularly as S. *aurantiacum*. The latter was found even more resistant in vitro to amphotericin and itraconazole as compared to those molecularly identified as *P. boydii/S. apiospermum*. The authors felt that most clinical laboratories would not be able to perform this differentiation, so antifungal susceptibility testing could provide the correct MIC per species [2].

6.1.4 Paecilomyces spp.

There can be great variability in the in vitro susceptibilities of various species of *Paecilomyces* sp. Most clinical laboratories do not always speciate an isolate, but rather may call it a *Paecilomyces* spp. Data in the literature on large numbers of isolates is not readily available. In a study in Spain in 2005, 11 isolates of *P. lilacinus* and 10 of *P. variotii* were tested using broth microdilution. For the *P. lilacinus*, $MIC_{90\%}$ was high, ≥ 8 µg/ml for amphotericin, itraconazole, and voriconazole. The $MIC_{90\%}$ for posaconazole indicated good activity at 0.5 µg/ml. The MICs for the *P. variotii* were all very low, indicating in vitro susceptibility to amphotericin, itraconazole, voriconazole, and posaconazole [12, 13]. There are reported differences between these two species vs. the echinocandins. *P. lilacinus* appears to be resistant to echinocandins with $MIC_{90\%} > 16$ µg/ml, *P. variotii* demonstrates very low MICs to anidulafungin and micafungin (≤ 0.03 µg/ml), but the $MIC_{90\%}$ of 17 isolates of *P. variotii* was found to be 4.0 µg/ml, which could be interpreted as resistant according to CLSI M38-A [9]. A study in Spain in 2008 compared the susceptibility of 27 *P. lilacinus* vs. 31 *P. variotii* that were identified morphologically and confirmed by molecular identification methods. *P. lilacinus* showed high MICs with geometric means >8 µg/ml for amphotericin, itraconazole, and all three echinocandins, whereas voriconazole, posaconazole, and terbinafine had geometric means <2 µg/ml with posaconazole demonstrating the best overall activity. *P. variotii* had low MICs and MECs (for echinocandins) for amphotericin, itraconazole, posaconazole, terbinafine, and the three echinocandins, but MICs were ≥ 2 μg/ml for 82% of the strains vs. voriconazole [8]. Identification and susceptibility testing would be recommended if treatment of a serious infection with *Paecilomyces* sp. was being considered [13].

6.1.5   Penicillium spp.

Determining the clinical significance of the isolation of a *Penicillium* spp. in a clinical specimen is difficult. Finding data on susceptibility testing in the literature is also sparse. A report from 35 isolates tested in the USA and Canada in 2000–2001 had favorable results of MIC or $MEC_{90\%}$ that were below the resistant breakpoint for amphotericin, caspofungin, and the azoles, itraconazole, voriconazole, and posaconazole; the percent of total isolates that had MIC < 1 μg/ml (*S*) was >77%, and for caspofungin, in particular, it was 97% [14]. In contrast, the Cuenca-Estrella study in 2006 reported on 45 *Penicillium* spp. from Spanish laboratories. The $MIC_{90\%}$ of amphotericin, itraconazole, and voriconazole was high and indicated that many strains were probably not susceptible. The $MIC_{90\%}$ to posaconazole was 2.0 μg/ml, which falls below the 4 μg/ml breakpoint often used to suggest susceptibility among the filamentous moulds [12]. Another study from Spain looked at MICs of 72 *Penicillium* spp., and the $MIC_{90\%}$ was high for itraconazole and voriconazole, but in this study, the posaconazole was also higher at 4.0 μg/ml. Results vs. micafungin and anidulafungin showed low $MIC_{90\%}$ at $\leq$0.06 μg/ml, but the caspofungin $MIC_{90\%}$ was 8.0 μg/ml which would indicate variability among the echinocandins [13]. Rarely is *Penicillium* spp. considered significant, but if it is, a susceptibility would be warranted to detect resistances since they do not appear to be predictable for all antifungal agents, nor the same when the organism is isolated from different countries.

6.1.6   Scopulariopsis spp.

Scopulariopsis brevicaulis is the most common species in this genus. It is most often involved in nondermatophyte nail infections, although there are rare reports of more systemic infections caused by this mould in recent decades. Nineteen isolates studied by broth microdilution were found nonsusceptible with very high $MIC_{90\%}$ to amphotericin (16 μg/ml), itraconazole, voriconazole, and posaconazole (>8 μg/ml) [12]. In another publication out of Spain specifically looking at combination testing vs. *S. brevicaulis*, 25 strains from nails, skin scrapings, sputum, and blood were found resistant to amphotericin and the azoles, and also highly resistant to terbinafine and caspofungin. Using a checkerboard MIC method for detecting results with combinations of amphotericin plus azoles or caspofungin, azoles plus caspofungin, or terbinafine plus azoles, synergy was noted for some strains especially with the combinations of posaconazole or voriconazole plus terbinafine. Most of the caspofungin combinations resulted in indifference, although if minimum effective concentrations (MECs) were used as the method of resulting, synergy was observed for more combinations including caspofungin plus amphotericin. Variability among strains may warrant synergy testing with individual patient isolates if treatment of systemic infections with *S. brevicaulis* is indicated [11].

6.1.7   Acremonium spp.

Acremonium spp. are a group of about 150 species of environmental isolates that rarely infect humans. Traumatic implantation of the spores of *Acremonium* spp. is the usual start of any infection with this group of hyaline moulds. A study at the Texas Fungus Testing Laboratory with 47 isolates from the USA demonstrated that all strains had very high MICs to amphotericin, the azoles, and echinocandins. Only terbinafine showed any activity with $MIC_{90\%} < 2.0$ µg/ml [28].

6.2   Zygomycetes

The $MIC_{90\%}$ of 15 *Rhizopus oryzae* isolates in Spain, using a broth microdilution, was 2 µg/ml for amphotericin, which is near what is considered a susceptible breakpoint, but it was ≥8 µg/ml for itraconazole, voriconazole, and posaconazole, suggesting probable resistance to the azoles. Sixteen other zygomycetes had similar $MIC_{90\%}$ results [12]. In 2009, 26 isolates of R. oryzae were tested using the EUCAST recommended methods for antifungal susceptibility testing, including a 24- and 48-h reading; the geometric mean vs. amphotericin was susceptible at 1.0 µg/ml. It was higher for itraconazole and voriconazole, indicating resistance to these azoles, although vs. posaconazole, the geometric mean at 48 h was only 2.0 µg/ml. Terbinafine testing resulted in very high MICs (32 µg/ml for the mean) [3]. Additionally, in another paper from Spain, 11 isolates of *R. oryzae* were found to have high $MIC_{90\%}$ to the three echinocandins (≥16 µg/ml). Other Zygomycetes, including 16 strains of *Myocladus corymbifereus*, 11 *Mucor* sp., and 11 other *Mucorales* sp. all demonstrated high $MIC_{90\%}$ to the azoles and echinocandins [13]. Twenty strains of *Mucor circinelloides* from a paper in Spain, using molecular methods for identification of the species, reported a low geometric mean for amphotericin, indicating susceptibility to the polyene, however, very high MICs to itraconazole, voriconazole, and posaconazole as well as terbinafine [3]. The antifungal susceptibility, using EUCAST proscribed methods, of 9 strains of *Saksenaea* spp. were reported in a paper from Europe and the USA. High MICs were demonstrated for amphotericin B and voriconazole, as were high MECs to the echinocandins. Low MICs were however shown for itraconazole, posaconazole, and terbinafine [5].

The Fungus Testing Laboratory in San Antonio, Texas, reported on susceptibilities of 217 Zygomycetes strains; isolates came from many US laboratories. One hundred percent of 134 isolates of *Rhizopus* spp. were susceptible to amphotericin; 0% to 5-FC, caspofungin, and fluconazole; >50% to itraconazole; >60% to posaconazole; and <5% to voriconazole. Results were similar for 41 *Mucor* spp.: 94% were susceptible to amphotericin, and 57% to itraconazole, and 70% to posaconazole. None were susceptible to 5-FC, caspofungin, fluconazole, or voriconazole. Twelve sp. of *Absidia* were susceptible to amphotericin and posaconazole, and >50% to itraconazole. In contrast, of the 13 species of *Cunninghamella* sp., only 63% were susceptible to amphotericin, 75% to posaconazole, and 29% to itraconazole [4].

Posaconazole in combination with amphotericin was found to perform synergistically in vitro against some strains of Zygomycetes in a study done in Germany and Spain and reported in 2008. Thirty clinical isolates of strains of *Rhizopus, Mucor, Absidia, Rhizomucor, Syncephalastrum*, and *Cunninghamella* were used in the study, and susceptibility testing was performed using a checkerboard method with preparations of both hyphae and conidia for each mould. The conidial preparation of two strains of *Rhizomucor* and one of *Rhizopus* demonstrated synergy with the combination; all others were indifferent. For the hyphal preparations, 12 strains (3 *Cunninghamella* sp., 23 *Mucor* sp., 3 *Rhizopus* sp., and 3 *Absidia* spp.) were synergistic, and the rest demonstrated indifference. No antagonism was seen with the combination of amphotericin and posaconazole [29].

6.3 Black (Dematiaceous) Fungi

6.3.1 *Alternaria spp.*

In one study from Spain, 11 *Alternaria* spp. had an $MIC_{90\%}$ of 0.5 μg/ml to amphotericin (probable susceptibility), and $MIC_{90\%} \geq 8$ μg/ml for itraconazole, voriconazole, and posaconazole [12]. In another paper, 11 *Alternaria alternata* and 10 *A. infectoria* had $MIC_{90\%}$ indicating high $MIC_{90\%}$ to amphotericin, the azoles, and the echinocandins, although there were some isolates with low MICs to amphotericin, so if considered significant in a patient when isolated, susceptibilities would be warranted. The $MIC_{90\%}$ of "other black fungi" was low, suggesting probable susceptibility to amphotericin and the echinocandins. The azoles remained in the resistant range [13]. In a paper from the Mayo Clinic in 2010, eight clinical isolates from five patients with *Alternaria* sp. were all found susceptible to itraconazole, posaconazole, amphotericin, and, caspofungin; they recommended itraconazole or an echinocandin for treatment. The MICs to voriconazole were higher than seen with the other azoles tested [6].

6.3.2 *Curvularia spp.*

Curvularia sp. can be a laboratory contaminant or rarely, a pathogen. A review of antifungal susceptibility testing literature in 2001 by Espinel-Ingroff et al. reported on susceptibilities vs. 26 isolates of *Curvularia* spp. (19 *C. lunata*, 3 each *C. verruculosa* and *C. senegalensis*, and 11 *C. inaequalis*). The range of MICs of *C. lunata* was 0.125 μg/ml to >16 μg/ml vs. amphotericin B; however, the MIC$_{90\%}$ was 0.5 μg/ml, which is interpreted as susceptible; the other species all have low MICs to amphotericin. Voriconazole was more active than itraconazole, but most *Curvularia* spp. were susceptible to both [15].

6.3.3   Exophiala spp.

One hundred sixty isolates of 11 different species of *Exophiala*, including 27 *E. dermatitidis*, 8 *E. jeanselmei*, 40 *E. oligosperma*, and 39 *E. xenobiotica* were tested in the Fungus Testing Laboratory, San Antonio, Texas, using a macrobroth dilution and following guidelines of CLSI M38-A. The MICs for three azoles, itraconazole, voriconazole, and posaconazole were low and in the susceptible range. All species except *E. attenuata* appeared susceptible to amphotericin B [19]. In a study of 16 *Exophiala dermatitidis* isolates in China, using a microbroth MIC method, the azoles and amphotericin were found to have low MICs in the susceptible range. MICs to caspofungin were high (32–64 μg/ml), and MICs of terbinafine were all ≤0.25 μg/ml. Synergism was demonstrated with the combination of caspofungin and voriconazole (10/16 isolates), caspofungin and amphotericin B (15/16 isolates), and caspofungin and itraconazole (16/1 isolates). The combination of terbinafine plus itraconazole showed neither synergism nor antagonism [36].

6.3.4   Other Dematiaceous Moulds

The geometric mean of terbinafine, amphotericin, and itraconazole when tested alone and results of combination testing of these agents against 53 isolates of black moulds in China were reported in 2008. For 22 strains of *Cladophialophora carrionii*, the mean MIC of amphotericin was 3.03 μg/ml (above the breakpoint of 1.0 μg/ml), for itraconazole it was 0.32 μg/ml (Susceptible), and for terbinafine it was 0.019 μg/ml (no breakpoints established for terbinafine). When terbinafine and itraconazole were combined, only 1/22 isolates of *C. carrionii* demonstrated synergism; no isolates demonstrated synergism or antagonism when terbinafine and itraconazole or amphotericin plus terbinafine were combined. With the 20 *Phialophora verrucosa* strains, the mean for amphotericin was 3.34 μg/ml (probably "intermediate"); for itraconazole, 0.69 μg/ml (susceptible); and for terbinafine, 0.07 μg/ml. For 11 *Fonsecaea pedrosoi*, the amphotericin mean MIC was 2.2 μg/ml (intermediate); for itraconazole, 0.43 μg/ml (susceptible); and for terbinafine, 0.05 μg/ml. Neither synergism nor antagonism was seen with the any of the combination of these three antifungal agents against the *Phialophora* or *Fonsecaea* [37].

   There was a review of antifungal susceptibility results of many yeast and moulds including dematiaceous moulds vs. amphotericin B, voriconazole, and itraconazole.

Table 6.1 Moulds (common in vitro patterns of susceptibility)

	Ampho	Itra	Vori	Posa	Echino
Hyaline Moulds					
Aspergillus					
A. fumigatus spp.	S	S	S	S	S
A. flavus	S	S	S	S	S
A. niger	S	V	V	S	S
A. nidulans	S	V, esp. between countries	V, esp. between countries		
A. terreus[a]	Usually R[b]	S	V	S	S
A. versicolor	Usually S	V	USA=S; Spain=V	S	S
Fusarium spp.					
F. solani	V	Non-S	S	V	Non-S
F. oxysporum	S	Non-S	V		Usually non-S
F. proliferatum	Non-S	Non-S	Non-R	Non-R	Usually non-S
Scedosporium sp.					
P. boydii	Intrinsic R	Non-S	Usually non-S	V	Non-S
S. prolificans	Intrinsic R	Non-S	Non-S	Non-S	Non-S
Penicillium spp. Spain	Non-S	Non-S	Non-S	V	A, M=S; C=V
Penicillium spp. USA/Canada	S	S	S	S	C=S
Paecilomyces sp.					
P. lilacinus[a]	Non-S	Non-S	V	S	Non-S
P. variotii	S	S	Usually S	S	S (A, M)
S. brevicaulis	Non-S	Non-S	Non-S	Non-S	Non-S
Zygomycetes					
Rhizopus oryzae	V	Non-S	Non-S	V	R
Rhizopus spp.[b]	S	V	R	V	Non-S
Mucor spp.	S[b]	V	R	V	R

	Ampho	Itra	Vori	Posa	Echino
Absidia spp.	S	V	Non-S	S	Non-S
Cunninghamella sp.	V	Usually Non-S	Non-S	Usually S	Non-S
Saksenaea[a]	Non-S	S	Non-S	S	Non-S
Dematiaceous moulds					
Alternaria sp.	S	V	V	S	V
Curvularia spp.	Usually S	S	S	I	I
Exophiala sp.[a]	Mostly S	S	S	S	R
Bipolaris sp.	S	V	V	I	I
Phialophora verrucosa[a]	Usually higher MICs	S	I	I	I
Fonsecaea[a]	S	S	I	I	I
Cladophialophora[a]	High MICs	S	I	I	I

S susceptible, Non-S MICs high and isolates usually considered not susceptible, R most often resistant, V variable results of in vitro testing, I insufficient data available to comment

Ampho amphotericin B, Flucon fluconazole, Itra itraconazole, Vori voriconazole, Posa posaconazole, Echino echinocandins, A anidulafungin, M micafungin, C caspofungin

[a]Terbinafine may have some in vitro activity

[b]Differences between USA/Canada and Spain

Fifty-eight isolates of *Bipolaris* sp. (23 *B. hawaiiensis*, and 32 *B. spicifera*, and 3 *B. australiensis*) had low MICs to amphotericin (≤1.4 µg/ml); *B. hawaiiensis* and *B. australiensis* had low MICs to itraconazole (<0.2 µg/ml), but MIC$_{90\%}$ for *B. spicifera* was 5.71 µg/ml, which would be considered resistant. Likewise with voriconazole, *B. spicifera* had higher MICs than *B. hawaiiensis* and *B. australiensis*, although all had MICs ≤ 2 µg/ml [15].

References

1. Alastruey-Izquierdo A, Cuenca-Estrella M, Monzon A et al (2008) Antifungal susceptibility of clinical *Fusarium* spp. isolates identified by molecular methods. J Antimicrob Chemother 61:805–809
2. Alastruey-Izquierdo A, Cuenca-Estrella M, Monzon A, Rodriguez-Tudela JL (2007) Prevalence and susceptibility of new species of *Pseudallescheria* and *Scedosporium* in a collection of clinical mould isolates. Antimicrob Agents Chemother 51:748–751
3. Alastruey-Izquierdo A, Castelli MA, Cuesta I et al (2009) Activity of posaconazole and other antifungal agents against *Mucorales* strains identified by sequencing of internal transcribed spacers. Antimicrob Agents Chemother 53:1686–1689
4. Almyroudis NG, Sutton DA, Fothergill AW et al (2007) In vitro susceptibilities of 217 isolates of zygomycetes to conventional and new antifungal agents. Antimicrob Agents Chemother 51:2587–2590
5. Alvarez E, Garcia-Hermoso D, Sutton DA et al (2010) Molecular phylogeny and proposal of two new species of the emerging pathogenic fungus *Saksenaea*. J Clin Microbiol 48:4410–4416
6. Boyce RD, Deziel PJ, Otley CC et al (2010) Phaeohyphomycosis due to *Alternaria* species in transplant recipients. Transpl Infect Dis 12:242–250
7. Bueid A, Howard SJ, Moore CB et al (2010) Azole antifungal resistance in *Aspergillus fumigatus*: 2008 and 2009. J Antimicrob Chemother 65:2116–2118
8. Castelli MV, Alastruey-Izquierdo A, Cuesta I et al (2008) Susceptibility testing and molecular classification of *Paecilomyces* spp. Antimicrob Agents Chemother 52:2926–2928
9. CLSI document M38A-2 (2008a) Reference method for broth dilution antifungal susceptibility testing of filamentous fungi: approved standard, 2nd edn. Clinical and Laboratory Standards Institute, Wayne, PA
10. Cuenca-Estrella M, Alastruey-Izquierdo A, Alcazar-Fuoli L et al (2008) In vitro activities of 35 double combinations of antifungal agents against *Scedosporium apiospermum* and *Scedosporium prolificans*. Antimicrob Agents Chemother 52:1136–1139
11. Cuenca-Estrella M, Gomez-Lopez A, Buitrago MJ et al (2006) In vitro activities of 10 combinations of antifungal agents against the multiresistant pathogen *Scopulariopsis brevicaulis*. Antimicrob Agents Chemother 50:2248–2250
12. Cuenca-Estrella M, Gomez-Lopez A, Mellado E et al (2006) Head-to-head comparison of the activities of currently available antifungal agents against 3378 Spanish clinical isolates of yeasts and filamentous fungi. Antimicrob Agents Chemother 50:917–921
13. Cuenca-Estrella M, Gomez-Lopez A, Mellado E et al (2009) Activity profile in vitro of micafungin against Spanish clinical isolates of common and emerging species of yeasts and molds. Antimicrob Agents Chemother 53:2192–2195
14. Diekema DJ, Messer SA, Hollis RJ et al (2003) Activities of caspofungin, itraconazole, posaconazole, revuconazole, voriconazole, and amphotericin B against 448 recent clinical isolates of filamentous fungi. J Clin Microbiol 41:3623–3626
15. Espinel-Ingroff A, Boyle K, Sheehan DJ (2001) In vitro antifungal activities of voriconazole and reference agents as determined by NCCLS methods: review of the literature. Mycopathologia 150:101–115

16. Espinel-Ingroff A, Johnson E, Hockey H, Troke P (2008) Activities of voriconazole, itraconazole and amphotericin B in vitro against 590 moulds from 323 patients in the voriconazole Phase III clinical studies. J Antimicrob Chemother 61:616–620

17. Espinel-Ingroff A, Diekema DJ, Fothergill A et al (2010) Wild-type MIC distributions and epidemiological cutoff values for the triazoles and six *Aspergillus* spp. for the CLSI broth microdilution method (M38-A2 Document). J Clin Microbiol 48:3251–3257

18. Fera MT, Lacamera E, DeSarro A (2009) New triazoles and echinocandins: mode of action, in vitro activity and mechanisms of resistance. Expert Rev Anti Infect Ther 7:981–998

19. Fothergill AW, Rinaldi MG, Sutton DA (2009) Antifungal susceptibility testing of *Exophiala* spp. a head-to-head comparison of amphotericin B, itraconazole, posaconazole, and voriconazole. Med Mycol 47:41–43

20. Gilgado F, Serena C, Cano J et al (2006) Antifungal susceptibilities of the species of the *Pseudallescheria boydii* complex. Antimicrob Agents Chemother 50:4211–4213

21. Gonzalez GM, Fothergill AW, Sutton DA et al (2005) In vitro activities of new and established triazoles against opportunistic filamentous and dimorphic fungi. Med Mycol 43:281–284

22. Heyn K, Tredup A, Salvenmoser S et al (2005) Effect of voriconazole combined with micafungin against *Candida, Aspergillus*, and *Scedosporium* spp. and *Fusarium solani*. Antimicrob Agents Chemother 49:5157–5159

23. Iqbal NJ, Boey A, Park BJ, Brandt ME (2008) Determination of in vitro susceptibility of ocular Fusarium spp. isolates from keratitis cases and comparison of Clinical and Laboratory Standards Institute M38-A2 and E test methods. Diagn Microbiol Infect Dis 62:348–350

24. Lass-Florl C, Alastruey-Izquierdo A, Cuenca-Estrella M et al (2009) In vitro activities of various antifungal drugs against *Aspergillus terreus*: global assessment using the methodology of the European Committee on Antimicrobial Susceptibility testing. Antimicrob Agents Chemother 53:794–795

25. Lewis RE, Wiederhold NP, Klepser ME (2005) In vitro pharmacodynamics of amphotericin B, itraconazole, and voriconazole against Aspergillus, Fusarium, and Scedosporium spp. Antimicrob Agents Chemother 49:945–951

26. Linares MJ, Charriel G, Solis F et al (2005) Susceptibility of filamentous fungi to voriconazole tested by two microdilution methods. J Clin Microbiol 43:250–253

27. Messer A, Moet GJ, Kirby JT et al (2009) Activity of contemporary antifungal agents, including the novel echinocandin, anidulafungin, tested against *Candida* spp., *Cryptococcus* spp., and *Aspergillus* spp.: report from the SENTRY antimicrobial surveillance program (2006 to 2007). J Clin Microbiol 47:1942–1946

28. Perdomo H, Sutton DA, Garcia D et al (2011) Spectrum of clinically relevant *Acremonium* species in the United States. J Clin Microbiol 49:243–256

29. Perkhofer S, Locher M, Cuenca-Estrella M et al (2008) Posaconazole enhances the activity of amphotericin B against hyphae of zygomycetes in vitro. Antimicrob Agents Chemother 52:2636–2638

30. Pfaller MA, Boyken L, Hollis RJ et al (2009) In vitro susceptibility of clinical isolates of *Aspergillus* spp. to anidulafungin, caspofungin, and micafungin: a head-to-head comparison using the CLSI M38-A2 broth microdilution method. J Clin Microbiol 47:3323–3325

31. Pfaller M, Boyken L, Hollis R et al (2011) Comparison of the broth microdilution methods of the European Committee on Antimicrobial Susceptibility Testing and the Clinical and Laboratory Standards Institute for testing itraconazole, posaconazole, and voriconazole against *Aspergillus* isolates. J Clin Microbiol 49:1110–1112

32. Pfaller MA, Castanheira M, Messer SA (2011) Echinocandin and triazole antifungal susceptibility profiles for *Candida* spp., *Cryptococcus neoformans*, and *Aspergillus* fumigatus: application of new CLSI clinical breakpoints and epidemiologic cutoff values to characterize resistance in the SENTRY antimicrobial surveillance program (2009). Diagn Microbiol Infect Dis 69:45–50

33. Pfaller MA, Messer SA, Boyken L et al (2008) In vitro survey of triazole cross-resistance among more than 700 clinical isolates of *Aspergillus* species. J Clin Microbiol 46:2568–2572

34. Pfaller M, Boyken L, Hollis R et al (2011) Use of epidemiological cutoff values to examine 9-year trends in susceptibility of *Aspergillus* species to the triazoles. J Clin Microbiol 49:586–590
35. Rodriguez-Tudela JL, Alcazar-Fuoli L, Mellado E et al (2008) Epidemiological cutoffs and cross-resistance to azole drugs in *A. fumigatus*. Antimicrob Agents Chemother 52:2468–2472
36. Sun Y, Liu W, Wan Z et al (2011) Antifungal activity of antifungal drugs, as well as drug combinations against *Exophiala dermatitidis*. Mycopathologia 171:111–117
37. Yu J, Li R, Zhang M et al (2008) In vitro interaction of terbinafine with itraconazole and amphotericin against fungi causing chromoblastomycosis in China. Med Mycol 46:745–747
38. Yustes C, Guarro J (2005) In vitro synergistic interaction between amphotericin B and micafungin against *Scedosporium* spp. Antimicrob Agents Chemother 49:3498–3500

Chapter 7
Usual Susceptibility Patterns for Systemic Dimorphic Fungi

Gerri S. Hall

Abstract There are not as many requests made for in vitro susceptibility testing of systemic fungi such as *H. capsulatum* and *B. dermatitidis*. There is limited data in the literature as well. The inoculum for testing could be the yeast or conidia/mycelial form of each of these fungi. Very few laboratories offer the susceptibility test. There is no standardized method in CLSI for testing of the dimorphic fungi. Below is some of the data found in the literature for in vitro testing of systemic dimorphic fungi when it has been done.

7.1 *Histoplasma capsulatum*

Four strains of *H. capsulatum* were tested using the mould and yeast form of the fungus vs micafungin, amphotericin, itraconazole, and fluconazole using a microdilution broth method of testing. The MICs vs amphotericin B were comparable when either the yeast or mould form of *H. capsulatum* was used and were all ≤0.5 µg/ml. The MICs vs itraconazole was very low (≤0.03 µg/ml) and comparable between yeast and mould inocula. The yeast form produced lower MICs to fluconazole vs the mould form. There was a huge difference between the yeast and mould forms of inocula when micafungin was tested. Very low MICs of ≤0.06 µg/ml were found vs the mould form, but >64 µg/ml to micafungin vs the yeast for of *H. capsulatum*. This was similar to what was seen with *B. dermatitidis* and *P. brasiliensis* [6].

The results of testing 144 clinical isolates of *H. capsulatum* vs amphotericin and three azoles were reported by Espinel-Ingroff in 2001. The range of MICs was ≤0.3–2.0 µg/ml with a MIC$_{90\%}$ of 0.25 µg/ml. The MIC$_{90\%}$ for 136 strains vs amphotericin was 1.0 µg/ml and vs. itraconazole, the MIC$_{90\%}$ was 0.06 µg/ml. The range

G.S. Hall, Ph.D. (✉)
Section of Clinical Microbiology, Department of Clinical Pathology,
Cleveland Clinic, Cleveland, OH 44195, USA
e-mail: hallg@ccf.org

G.S. Hall (ed.), *Interactions of Yeasts, Moulds, and Antifungal Agents:*
How to Detect Resistance, DOI 10.1007/978-1-59745-134-5_7,
© Springer Science+Business Media, LLC 2012

of MICs for 26 isolates vs fluconazole was ≤0.125–≥64 µg/ml [3]. The same author in a subsequent chapter summarized the echinocandin susceptibility of the systemic dimorphic fungi from a number of literature reports. Using the mould phase of five isolates of *H. capsulatum*, the MICs vs caspofungin was 0.5–4 µg/ml; the same five strains vs anidulafungin demonstrated an MIC range of 2–4 µg/ml [1].

7.2 *Blastomyces dermatitidis*

Six strains of *B. dermatitidis* (three ATTC strains and three clinical strains) were tested using the mould and yeast form of the fungus vs micafungin, amphotericin, itraconazole, and fluconazole using a microdilution broth method of testing. The MICs vs amphotericin B were different when the yeast vs mould form of *B. dermatitidis* was used, i.e., the MIC with the mould form was 0.0156 µg/ml and the yeast for all was 0.125 µg/ml. The MICs vs itraconazole were very low (≤0.03 µg/ml), comparable between yeast and mould inocula and even lower than found vs *H. capsulatum* for most strains. Similar to testing of *H. capsulatum*, the yeast form produced lower MICs to fluconazole vs the mould form and there was a huge difference between the yeast and mould forms of inocula when micafungin was tested. Very low MICs of ≤0.03 µg/ml were found vs the mould form, but was ≥32 µg/ml to micafungin vs the yeast for of *B. dermatitidis*. This was similar to what was seen with *H. capsulatum* and *P. brasiliensis* [6].

The same 2001 paper described above for *H. capsulatum* reported on results of 142 isolates of *B. dermatitidis* vs. voriconazole. MIC$_{90\%}$ was 0.25 µg/ml. One hundred thirty three isolates vs amphotericin had MIC$_{90\%}$ of 0.5 µg/ml and vs itraconazole MIC$_{90\%}$ was 0.125 µg/ml. The range of MICs for 30 isolates vs fluconazole was 1–64 µg/ml [2, 3].

Another study with 34 isolates of *B. dermatitidis* provided results on the echinocandins: the MIC range was 2–>8 µg/ml for anidulafungin, and for fewer isolates, the range for caspofungin and micafungin was 0.5–>8 µg/ml [2]. In the 2003 literature review paper, the azoles, itraconazole, and posaconazole appeared to be more active than the echinocandins [1].

7.3 *Coccidioides immitis*

In one study looking at four clinical isolates of *C. immitis* (along with other dimorphic fungi as noted above), using, of course, only the mould form of this fungus, very low MICs were demonstrated vs amphotericin (<0.25 µg/ml), itraconazole (≤0.125 µg/ml), fluconazole (4 µg/ml), and micafungin (0.02 µg/ml) [6].

For *C. immitis*, the 2001 paper from Espinel-Ingroff reported on 142 isolates vs voriconazole with MIC range of ≤0.3–0.5 µg/ml and MIC$_{90\%}$ of 0.25 µg/ml. The amphotericin and itraconazole MIC$_{90\%}$ for 131 isolates was 1.0 µg/ml for both agents. The range of MIC for fluconazole vs 29 *C. immitis* isolates was 2–64 µg/ml [3].

Results were similar in a later paper of 29 isolates vs amphotericin and fluconazole, but echinocandins were also reported. For caspofungin, 25 C. *immitis* isolates had MIC of >8 μg/ml; there were results for only four isolates vs micafungin, but the MIC was in the susceptible range of 0.01 μg/ml interestingly [1].

7.4 *Paracoccidioides brasiliensis*

Seven clinical strains of *P. brasiliensis* were tested using the mould and yeast form of the fungus vs micafungin, amphotericin, itraconazole, and fluconazole using a microdilution broth method of testing. The MICs vs amphotericin B were different when the yeast vs mould form of *P. brasiliensis* was used, i.e., the MIC with the mould form was 0.03 μg/ml and the yeast for all was ≤0.25 μg/ml. The MICs vs itraconazole were very low (≤0.08 μg/ml), comparable between yeast and mould inocula and even lower than found in vs *H. capsulatum* for most strains. The yeast form of *P. brasiliensis* produced slightly lower MICs to fluconazole vs the mould form for 3/7 strains. The difference between the yeast and mould forms of inocula when micafungin was tested was present as seen with *B. dermatitidis* and *H. capsulatum* although both were quite high (>4 μg/ml) [6].

Only 19 isolates of *P. brasiliensis* were reported in the 2001 paper by Espinel-Ingroff vs voriconazole. The range of MIC values was ≤0.03–2 μg/ml; for amphotericin, ≤0.125 to >64 μg/ml; for itraconazole, ≤ 0.03–1 μg/ml; and for fluconazole, ≤0.125–64 μg/ml [3].

7.5 *Sporothrix schenckii*

Seven clinical strains of *S. schenckii* were tested using the mould and yeast form of the fungus vs micafungin, amphotericin, itraconazole, and fluconazole using a microdilution broth method of testing. The MICs vs amphotericin B and itraconazole were all ≤2.0 μg/ml, which was slightly higher than the MICs for the other dimorphics tested. The MICs to fluconazole were very high (≥16 μg/ml) and for most >64 μg/ml. There were results of both yeast and mould phase inocula for only 2/7 strains of *S. schenckii*, and for those two, the mould form produced a lower MIC to micafungin (<1.0 μg/ml) as was seen with the other dimorphics. With yeast inocula, ≥16 μg/ml of micafungin was found [5]. Espinel-Ingroff reported that for five clinical isolates of *S. schenckii*, the range of MIC to the echinocandins caspofungin and anidulafungin was broad, from 0.25–>8 μg/ml. The MIC range for posaconazole for those same isolates was 0.12–1.0 μg/ml [2].

In a Brazilian study in 2006, 43 isolates of *S. schenckii* were tested using CLSI standards M27-A2. Higher MICs were noted with the mould form of inocula vs amphotericin, itraconazole, and terbinafine, especially vs amphotericin B. This was in contrast to the higher MICs seen with yeast form in the Japanese study by Nakai. [5, 6] Ninety-five percent of the results using yeast form were susceptible to amphotericin B (MIC ≤1.0 μg/ml) vs 35% using the mycelial form as inocula. For itraconazole, 98% of the isolates had an MIC below 0.5 μg/ml with yeast form vs 76% with mould form of inocula; for terbinafine, the MICs to which most of the isolates were susceptible was 0.06–0.25 μg/ml with minimal variability between the mould vs yeast forms [5]. Ninety one *S. schenckii* isolates from another Brazilian study in 2009 also demonstrated higher MICs with mycelial forms as the inocula compared to the yeast form although good activity was seen with either mycelial or yeast forms vs. amphotericin, itraconazole, terbinafine, and ravuconazole; no activity was demonstrated against *S. schenckii* when either growth phase was used against fluconazole and voriconazole. When an E-test method was compared to the broth microdilution, many discrepancies were seen with voriconazole and itraconazole, but the authors concluded that the E-test could be employed instead of the broth microdilution method for testing amphotericin and fluconazole [4].

The voriconazole MIC$_{90\%}$ for 47 isolates of *S. schenckii* was >16 μg/ml; for itraconazole, 4.0 μg/ml; and for fluconazole, >128 μg/ml. The MIC$_{90\%}$ for amphotericin was 4.0 μg/ml [3]. Five isolates were reported vs echinocandins, and the MIC range was 0.25–>8 μg/ml. Posaconazole results were included as well, 0.12–1.0 μg/ml [2, 3].

7.6 *Penicillium marneffei*

Five clinical strains of *P. marneffei* were tested in the Nakai paper along with the other dimorphics. MICs for both yeast and mould inocula vs amphotericin were similar and low (≤0.5 μg/ml), itraconazole (≤0.06 μg/ml), and fluconazole (≤4 μg/ml). MICs to micafungin were >4 μg/ml with the yeast inocula. For three out of four strains of *P. marneffei*, the mould inocula gave an MIC of 2.0 μg/ml and for 1 strain 0.03 μg/ml [6].

In another study, 27 isolates had an MIC to voriconazole of ≤0.03 μg/ml, and 25 isolates were susceptible to fluconazole with MICs ≤8 μg/ml. The range for 25 isolates vs amphotericin was ≤0.03–8 μg/ml and for itraconazole, <0.03–2 μg/ml [3].

In summary, for all of the dimorphic fungi reported on in the literature, a variety of differences were seen with the growth phase used, i.e., mycelial vs yeast phase; there appear to also be differences seen when results are given for isolates tested from different countries. Knowing the in vitro susceptibility of isolates from the country in which you practice would be important when planning treatment programs for these fungi. Performance of in vitro susceptibility in cases of serious infections may be warranted.

References

1. Espinel-Ingroff A (2003) In vitro antifungal activities of anidulafungin and micafungin, licensed agents and the investigational triazole posaconazole as determined by NCCLS methods for 12,052 fungal isolates: review of the literature. Rev Iberoam Microbiol 20:121–136

2. Espinel-Ingroff A (1998) Comparison of in vitro activities of the new triazole SCH56592 and the echinocandins MK-0991 (L-743,872) and LY303366 against opportunistic filamentous and dimorphic fungi and yeasts. J Clin Microbiol 36:2950–2956
3. Espinel-Ingroff A, Boyle K, Sheehan DJ (2001) In vitro activities of voriconazole and reference agents as determined by NCCLS methods: review of the literature. Mycopathologia 150:101–115
4. Gutierrez-Galhardo MC, Zancope-Oliveira RM, Monzon A, Rodriguez-Tudela JL, Cuenca-Estrella M (2010) Antifungal susceptibility profile in vitro of Sporothrix schenckii in two growth phases and by two methods: microdilution and E-test. Mycoses 53:227–231
5. Kohler LM, Soares BM, Santos DA, Barros ME, Hamdan JA (2006) In vitro susceptibility of isolates of Sporothrix schenckii to amphotericin B, itraconazole, and terbinafine: comparison to yeast and mycelial forms. Can J Microbiol 52:843–847
6. Nakai T, Uno J, Ikeda F, Tawara S, Nishimura K, Miyaji M (2003) In vitro antifungal activity of micafungin (FK463) against dimorphic fungi: comparison of yeast-like and mycelial forms. Antimicrob Agents Chemother 47:1376–1381

Chapter 8
Utility of Antifungal Susceptibility Testing and Clinical Correlations

Daniel J. Diekema and Michael A. Pfaller

Abstract In this chapter, we review the available published data addressing the clinical relevance of antifungal susceptibility test results. By far the most data exist to support the clinical relevance of AFST results for *Candida* against fluconazole, and these data suggest that the clinical utility of this information mirrors that put forward for antibacterial susceptibility testing. Clinical relevance has also been demonstrated for selected other antifungal agents against *Candida* and *Cryptococcus* spp. By contrast, little direct support for the clinical utility of AFST for moulds is available.

8.1 Antimicrobial Susceptibility Testing and "Clinical Utility"

The first and most important step toward the goal of "clinically useful" antimicrobial susceptibility testing is the development of standardized, reproducible reference testing methods. For antifungal susceptibility testing (AFST), this step has been accomplished and has resulted in the establishment of standard testing methods by both the Clinical and Laboratory Standards Institute (CLSI) and the European Committee for Antimicrobial Susceptibility Testing (EUCAST) [1–6]. These groups have developed and validated methods for broth dilution (CLSI and EUCAST) and

D.J. Diekema, M.D. (✉)
Division of Infectious Diseases, Department of Internal Medicine, University of Iowa
Carver College of Medicine, Iowa City, IA, USA

University of Iowa College of Public Health, Iowa City, IA, USA
e-mail: daniel-diekema@uiowa.edu

M.A. Pfaller, M.D.
Division of Clinical Microbiology, Department of Pathology, University of Iowa
Carver College of Medicine, Iowa City, IA, USA

University of Iowa College of Public Health, Iowa City, IA, USA

G.S. Hall (ed.), *Interactions of Yeasts, Moulds, and Antifungal Agents:*
How to Detect Resistance, DOI 10.1007/978-1-59745-134-5_8,
© Springer Science+Business Media, LLC 2012

disk diffusion (CLSI) susceptibility testing of yeasts [1, 2, 5] and moulds [3, 4, 6] (see Chap. 2). Development of standard methods has been accompanied by development of quality control limits for many agents for both minimum inhibitory concentration (MIC) and disk diffusion (DD) methods [1–9].

Now that AFST tools are available, how should we evaluate and use them? Standard AFST methods have clearly been very useful for performing large-scale surveillance studies, describing the susceptibility profiles of fungal organisms worldwide, and for examining geographic, temporal, and species-related trends in vitro susceptibility [10–34]. Standard methods have also proven useful for screening investigational agents to determine the likelihood that they hold clinical promise. In addition, standard AFST methods can now serve as gold standard comparators for the development and validation of commercial AFST methods ([35–39], see Chap. 3). All of these uses, however, presume that the information provided by standard AFST methods (the MIC or the disk zone diameter) has clinical relevance in the care of patients with fungal infection. It is this issue (the "clinical utility" or "clinical relevance" of AFST) that will be the topic of this chapter.

Rex and Pfaller have articulated several important principles to consider when discussing the clinical utility of susceptibility test methods [40]. These principles include an understanding that the MIC is a construct that is largely defined by testing conditions, rather than a physical or chemical measurement (e.g., a sodium level). While it is hoped that this highly artificial measure will correlate with clinical outcome, several factors related to the host (immune response, underlying illness, site of infection), the infecting organism (virulence), and the antifungal agent (dose, pharmacokinetics, pharmacodynamics, drug interactions) may be more important than susceptibility test results in determining clinical outcomes for infected patients. In particular, in vitro susceptibility of an organism to an antifungal agent does not predict a successful therapeutic outcome. However, in vitro resistance of an organism to an antifungal agent should help predict clinical failure [40].

Understanding these principles should temper our expectations regarding the clinical utility of any in vitro susceptibility test and should engender a healthy respect for how difficult it is to demonstrate an association between an AFST result and clinical outcome for patients with serious fungal infections. Since AFST methods have been widely introduced more recently than have antibacterial susceptibility testing (ABST) methods, we will first review lessons learned from the experience to date with ABST.

8.2 Lessons Learned from Antibacterial Susceptibility Testing: The "90–60 Rule"

Rex and Pfaller reviewed 12 studies that sought to correlate in vitro susceptibility test results with clinical outcome among persons with bacterial infection [40]. While these studies, outlined in detail by Rex and Pfaller and summarized in Fig. 8.1,

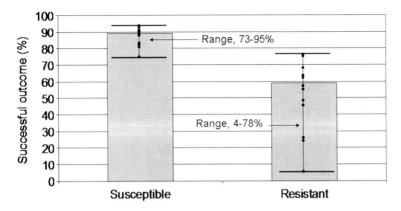

Fig. 8.1 Rates of successful outcome of therapy for organisms susceptible versus resistant to the therapeutic agent used (Data represent compilation of the results of 12 studies reviewed in Ref. [40])

represented a wide variety of infection sites, antimicrobial agents administered, and measures of success (mortality, clinical response, microbiologic response), a clear pattern emerged: infections due to susceptible bacteria responded favorably to appropriate therapy about 90% of the time, while infections due to resistant bacteria, when treated with the agent to which they were resistant, responded about 60% of the time [40].

This observation, referred to by the authors as the "90–60 rule," is best understood in the context of the principles of in vitro susceptibility testing described above. The fact that appropriate therapy fails in up to 10% or more of infected patients demonstrates the importance of host factors in clinical outcome. Likewise, the fact that 60% of patients treated with a drug to which their infecting organism is resistant still respond is a testament to the importance of host immune response, among other factors, in determining outcome. Antibacterial susceptibility testing appears to have its greatest clinical utility in predicting which antimicrobials are less likely to result in a favorable clinical outcome.

8.3   Clinical Correlations for Antifungal Susceptibility Testing

8.3.1   Candida *spp.*

By far the most in vivo data to support clinical relevance of AFST exist for the *Candida* species. As such, we will spend most of this chapter reviewing the data for *Candida* for each antifungal agent.

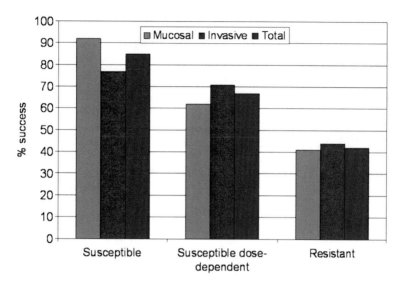

Fig. 8.2 Correlation of fluconazole susceptibility with clinical response for mucosal and invasive *Candida* infections treated with fluconazole (Adapted from Ref. [28]). $N = 1,295$ infection-episode-isolate events (692 mucosal, 603 invasive, compiled from 12 published studies)

8.3.1.1 Candida spp. and Fluconazole

The original CLSI interpretive breakpoints for fluconazole and *Candida* spp. (susceptible (S), MIC ≤8 μg/mL, susceptible dose dependent (S-DD), MIC 16–32 μg/ mL, and resistant (R), MIC ≥64 μg/mL) were based upon an analysis of treatment outcome in a dataset consisting of both mucosal (411 patient-episode-isolate events) and invasive (108 patient-episode-isolate) disease [41]. The interpretive category of "S-DD" was first introduced at this time and served to emphasize the importance of adequate fluconazole dosing to achieve blood and tissue levels sufficient to inhibit isolates with higher fluconazole MICs. Overall, the clinical response rate in this dataset was 87%, including 92% (370/403) for infections with susceptible isolates, 82% (45/55) for infections with isolates having MIC in the S-DD range, and 56% (34/61) for those infections with resistant isolates. When invasive infection episodes only were examined, the corresponding success rates were 71%, 91%, and 58%, respectively. As has been pointed out previously, the limitations of this dataset include (1) the majority of results were drawn from mucosal candidiasis episodes, (2) few episodes were due to isolates with elevated fluconazole MICs, and (3) the concept of dose-dependent susceptibility was demonstrated only for mucosal disease [42].

Since these data were published and breakpoints were established, there have been several additional studies examining the relationship between therapeutic outcome and fluconazole MIC or susceptibility category. These 12 studies, summarized in Fig. 8.2, include 692 mucosal and 603 invasive infection episodes. The overall success rate for the 1,295 patient-episode-isolate events was 77%, including

85% for episodes due to *Candida* with MIC ≤8 μg/mL (S), 67% for episodes with MIC of 16–32 μg/mL (S-DD), and 42% for those episodes due to resistant isolates (MIC ≥64 μg/mL). These data provide confirmation of the clinical relevance of antifungal susceptibility testing and are consistent with the "90–60 rule" described above.

As would be expected, a low MIC (≤8 μg/mL) to fluconazole was more predictive of success for patients with mucosal, rather than invasive, candidiasis. The patient population at risk for invasive candidiasis is such that host factors often overwhelm the in vitro potency of an antifungal therapy in determining the outcome of infection. In contrast, a high MIC (≥64 μg/mL) predicted a higher likelihood of failure in patients with both mucosal and invasive candidiasis (only ~40% in each group were successfully treated). Clearly, *Candida* isolates for which the fluconazole MIC is ≥64 μg/mL represent organisms for which adequate fluconazole concentrations cannot be sustained with current dosing regimens. These organisms cause infections that are significantly less likely to respond to fluconazole therapy. They also happen to consist mainly of *C. glabrata* and *C. krusei*.

The available data also support the concept that *Candida* spp. with MICs in the S-DD range have response rates comparable to susceptible isolates provided that higher fluconazole doses are used [42–46]. For both mucosal and invasive disease, response rates to fluconazole were higher for S-DD than for resistant organisms (Fig. 8.2).

Because physicians need useful susceptibility information sooner [47–50], the CLSI Subcommittee sought to determine if reading the BMD fluconazole MIC at 24 h would produce valid results when interpreted using the existing 48-h breakpoints [51]. Excellent correlation ($R=0.9$) was noted between fluconazole MICs read after 24- and 48-h incubation for 11,654 isolates of *Candida* spp. [52]. The categorical agreement between the two readings, using the 48-h breakpoints, was also very good, with an absolute categorical agreement of 93.8% and only 0.02% very major (false-susceptible) or major (false-resistant) errors. Furthermore, when the 48-h fluconazole breakpoints were applied to MICs read at 24 h, the earlier reading predicted therapeutic outcome as accurately as the 48-h MICs in a set of 528 isolates from patients with mucosal and invasive candidiasis: 82% success for episodes in which the 24-h MIC was ≤8 μg/mL (S), 55% for those episodes in which the MIC was 16–32 μg/mL (S-DD), and 39% for episodes in which the MIC was ≥64 μg/mL (R) [51]. Based on these results, the CLSI Subcommittee has included the option to read fluconazole MICs for *Candida* after 24-h incubation, using the original interpretive breakpoints, in CLSI document M27-A3 [1]. Physicians and laboratories are cautioned to be aware that when an isolate is identified as *C. glabrata* and the fluconazole MIC is ≤32 μg/mL (read at either 24 or 48 h), patients should receive a maximum dosage of fluconazole (e.g., 12 mg/kg/day) [1, 53].

Additional data in support of the CLSI breakpoints for fluconazole, and for the S-DD category, come from the pharmacodynamic literature. In vivo pharmacodynamic studies suggest that the parameter most predictive of efficacy in animal models is the area under the concentration curve (AUC) to MIC ratio (AUC/MIC) [43, 44, 54, 55]. In healthy adults with normal renal function, the AUC is approximated

Table 8.1 Relationship between dose/MIC ratio and clinical response in fluconazole treatment of mucosal and invasive candidiasis

Dose/MIC ratio	% Clinical success (n/N)[a, b]
≥400	98 (123/125)
100–300	95 (135/142)
50–75	86 (53/62
25–37.5	72 (178/246)
6.26–12.5	68 (52/77)
≤6.25	55 (36/65)

[a] n number of successful treatment events, N number of total episode-isolate events
[b] Data compiled from four studies, table adapted from Ref. [28]

by the total daily dose in milligrams [42]. Using these assumptions, the CLSI Subcommittee for Antifungal Testing used the dose/MIC ratio (as a surrogate for AUC/MIC) to analyze the relationship between drug dose, organism MIC, and clinical outcome for fluconazole treatment of mucosal candidiasis [42]. These data, outlined in Table 8.1, demonstrate that a dose/MIC ratio of >25 is most predictive of efficacy. This observation further supports the CLSI breakpoints for fluconazole against *Candida* spp.: a 400-mg/day dose will achieve a ratio ≥50 for organisms with fluconazole MIC ≤8 µg/mL, while 800 mg/day is required to achieve dose/MIC ratios of >25 for isolates with MICs of 16–32 µg/mL.

While most data for clinical relevance of fluconazole MIC and *Candida* spp. derives from the CLSI method [1], there is correlation between fluconazole MIC and dose/MIC ratio and clinical outcome using the EUCAST method, using cohorts of patients with mucosal candidiasis and candidemia [56]. These investigators reported a clinical success rate of 94% (136 of 145 infection episodes) when fluconazole MICs were ≤2 µg/mL, 66% (8 of 12 episodes) when the fluconazole MIC was 4 µg/mL, and only 12% (12 of 101 episodes) when the MIC was ≥8 µg/mL. However, over half of the patients in these cohorts were treated with fluconazole doses <400 mg/day. The clinical response rate for dose/MIC ratio of >25 was 91.2% (145 of 159 episodes) in this study [56]. A second report utilized data mining of these same two cohorts to validate the EUCAST breakpoints for *Candida* and fluconazole (susceptible (S), MIC ≤2 µg/mL; intermediate (I), MIC 4 µg/mL; and resistant (R), MIC ≥8 µg/mL) which differ from those of CLSI [57]. EUCAST breakpoints are species specific and apply only to *C. albicans*, *C. tropicalis*, and *C. parapsilosis*. Comparison of CLSI and EUCAST breakpoints was initially difficult, given the differences in MIC methodology and in the definitions used to assess clinical correlation [56].

Therefore, the CLSI Subcommittee on Antifungal Susceptibility Testing embarked upon an effort to harmonize the CLSI and EUCAST breakpoints, taking into account the wild-type distributions by species, molecular mechanisms of resistance, essential and categorical agreement of MICs generated using the two methods, pharmacokinetic and pharmacodynamic considerations, and a reexamination

Table 8.2 Analysis of fluconazole MIC ranges and outcome in patients with candidemia and mucosal candidiasis

MIC (μg/mL)	Outcome at MIC	
	No. of events	% Success
≤2	550	91.6
4	52	82.7
≥8	212	37.3

Adapted with permission from Ref. [58]

of the correlation between MICs and outcomes using previously published data [58]. Datasets correlating CLSI and EUCAST MICs with outcomes revealed lower response rates when MICs were >4 μg/mL for *C. albicans, C. tropicalis*, and *C. parapsilosis* and >16 μg/mL for *C. glabrata* (Table 8.2). These findings led CLSI to adjust their breakpoints for fluconazole and *C. albicans, C. tropicalis, C. parapsilosis* (S, ≤2 μg/mL; S-DD, 4 μg/mL; R, ≥8 μg/mL), and *C. glabrata* (S-DD, ≤32 μg/mL; R, ≥64 μg/mL). These breakpoints provide consistency with EUCAST breakpoints and should also be more sensitive for detection of emerging resistance among common *Candida* spp. [58].

In summary, abundant data exist to support the clinical relevance of AFST of fluconazole against *Candida* spp. However, some of these data may be confounded by the fact that most clinical *Candida* isolates that have elevated MICs to fluconazole are *C. glabrata* or *C. krusei*. If clinical variables exist that are associated with both increased risk for *C. glabrata* or *C. krusei* infection, and with poor clinical outcome, then the impact of the organisms MIC may be negligible. Data to support clinical relevance may then be found either by examining only infection episodes due to a single species or by examining case reports and series of salvage therapy, wherein patients infected with isolates with elevated fluconazole MICs who are failing fluconazole therapy are treated successfully with other antifungal agents to which the organism has lower MICs (e.g., caspofungin, voriconazole, amphotericin B) [58–60].

One obstacle to demonstrating clinical relevance within a single *Candida* species that is usually fluconazole-susceptible (e.g., *C. albicans, C. tropicalis, C. parapsilosis*) is the absence of sufficient numbers of isolates that are fluconazole-resistant. In order to establish a relationship between MIC and clinical outcome, one requires not only sufficient numbers of resistant isolates but also a sufficient number of patients infected with resistant isolates and *treated with the drug to which the isolate is resistant*. The same problem exists for newer drugs that have excellent in vitro activity against *Candida* spp. (e.g., echinocandins).

8.3.1.2   Candida spp. and Voriconazole

AFST interpretive breakpoints for voriconazole against *Candida* spp. were proposed [27] and adopted by CLSI in 2008 [1]. Like the fluconazole breakpoints, an

Table 8.3 *Candida* species, geometric mean MICs, and investigator-assessed response to voriconazole therapy

Species	No. of isolates tested	Geometric mean MIC (μg/mL)[a]	% Success
C. albicans	96	0.02	72
C. parapsilosis	34	0.03	85
Candida spp.	12	0.07	92
C. tropicalis	51	0.13	73
C. krusei	9	0.37	78
C. glabrata	47	0.79	55

Table adapted with permission from Ref. [27]
[a]Broth microdilution MICs were determined in accordance with CLSI M27-A2

Table 8.4 Investigator assessment of efficacy versus baseline MIC for *Candida* species against voriconazole in primary and salvage therapy studies

MIC breakpoint (μg/mL)	Interpretive category	No. of isolates	% Success
≤0.5	S	211	73
1–2	S-DD	17	65
≥4	R	21	62
≤1	S	221	74
2	S-DD	7	43
≥4	R	21	62

Adapted with permission from Ref. [27]

S-DD category was included (MIC for susceptible, ≤1 μg/mL; S-DD, 2 μg/mL; and resistant, ≥4 μg/mL). These breakpoints were established as they should be for any organism-drug combination, based upon an integration of the MIC distribution, pharmacokinetic and pharmacodynamic parameters, and the relationship between in vitro activity and outcome from in vivo and clinical studies [27]. Given the scope of this chapter, we will concentrate on the latter—the relationship between MIC and outcome from clinical studies.

Pfaller, et al. summarized the clinical trials data available from 249 patients infected with *Candida* spp. and treated with voriconazole [27]. As outlined in Table 8.3, the clinical efficacy of voriconazole was >72% for all species of *Candida* except for *C. glabrata*. For *C. glabrata*, the mean voriconazole MIC was higher (0.8 versus <0.4 μg/mL for the other species), and the clinical success rate was lower (55%). Moreover, analysis of the data demonstrated a statistically significant relationship between baseline MIC of the infecting *Candida* isolate and end-of-therapy assessment of outcome ($p = 0.02$) [27]. Because of the strong association with species identification and voriconazole MIC, species identification was also significantly associated with end-of-therapy assessment of outcome in this dataset, and when both variables were entered into a model, the log MIC term was no longer significant [27]. Table 8.4 reveals the percent success rate of therapy by investigator assessment of outcome versus the MIC category for two possible voriconazole breakpoint criteria. For this dataset, the resistant category (MIC ≥4 μg/mL) was

Table 8.5 Analysis of voriconazole MIC range versus outcome in patients with candidiasis

Species (no. tested)	Incubation time (h)	No. of events (% success) at MIC (µg/mL)		
		≤0.125	0.25–0.5	≥1
C. albicans (96)	24	83 (72.3)	6 (100.0)	7 (42.8)
	48	81 (74.1)	7 (71.4)	8 (50.0)
C. tropicalis (48)	24	47 (72.3)	1 (100.0)	
	48	35 (80.0)	7 (42.9)	6 (66.7)
C. parapsilosis (34)	24	34 (85.3)		
	48	34 (85.3)		
Miscellaneous (12)	24	9 (88.9)	2 (100.0)	1 (0.0)
	48	9 (88.9)	1 (100.0)	2 (50.0)
Total (190)	24	173 (75.7)	9 (100.0)	8 (37.5)
	48	159 (78.6)	15 (60.0)	16 (56.3)

Adapted with permission from Ref. [62]

associated with an approximately 60% success rate, while infections due to susceptible organisms responded approximately 75% of the time [27].

An S-DD category could be supported for voriconazole based upon the same principles upon which it was established for fluconazole, in order to account for the nonlinear pharmacokinetics and the dosing flexibility of the drug [41]. Moreover, the S-DD or "intermediate" (I) categories can function as a "buffer zone" to prevent minor technical factors from causing discrepancies that cross major interpretive categories.

There are also EUCAST breakpoints for voriconazole and *C. albicans*, *C. tropicalis*, and *C. parapsilosis*, which differed from those of CLSI [61]. Recently, however, a reassessment of the CLSI breakpoints was performed in an effort to harmonize the two methods [62]. After establishing excellent agreement between 24-h MICs using CLSI and EUCAST methods, and after examining the correlation between MICs and outcomes from previously published data using CLSI methods (see Table 8.5), the subcommittee recommended adjusted breakpoints for *C. albicans*, *C. tropicalis*, *C. parapsilosis*, and *C. krusei* (Table 8.6).

8.3.1.3 Candida spp. and Itraconazole or Posaconazole

AFST interpretive breakpoints for itraconazole against *Candida* spp. have also been established by CLSI, but are limited by virtue of (1) being based only upon episodes of mucosal disease and (2) being established prior to the availability of intravenous preparations of itraconazole, and therefore subject to the limitations of erratic bioavailability of the oral preparation [41]. The current breakpoints (susceptible, ≤0.125 µg/mL; S-DD, 0.25–0.5 µg/mL; and resistant, ≥1 µg/mL) are therefore conservative and not readily comparable to those for fluconazole or for those likely to be approved for newer azoles. Given the limited use of itraconazole for invasive candidiasis, there are no additional large datasets available, as there are for fluconazole, to provide additional support for the existing breakpoints.

Table 8.6 New 24-h CLSI MIC interpretive breakpoints for voriconazole and *Candida*[a]

Species	MIC (μg/mL) defining		
	Susceptible	Intermediate	Resistant
C. albicans	≤0.125	0.25–0.5	≥1
C. tropicalis	≤0.125	0.25–0.5	≥1
C. parapsilosis	≤0.125	0.25–0.5	≥1
C. krusei	≤0.5	1	≥2
C. glabrata	Note[b]		

[a]Breakpoints may also be used for 48-h readings if 24-h growth control is insufficient
[b]*Note*: The current data is insufficient to demonstrate a correlation between in vitro susceptibility testing and clinical outcome for *C. glabrata* and voriconazole. The epidemiological cutoff values (MIC, ≤0.5 μg/mL) may be used to differentiate wild-type (WT) from non-WT (strains with acquired or mutational resistance mechanisms) strains of this species

Regarding posaconazole, at the time of this writing, there are no clinical breakpoints, and no data are available to support the clinical relevance of posaconazole MICs for treatment of *Candida* spp. infections.

8.3.1.4 Candida spp. and Flucytosine

CLSI established breakpoints for flucytosine tested against *Candida* spp. [1, 41], but these were based primarily upon historical and animal model data. No data exists that directly addresses the clinical relevance of AFST results for flucytosine. Since flucytosine should not be used as a single agent, but only in combination with other systemic antifungal drugs such as amphotericin B or fluconazole, establishing clinical relevance for AFST of flucytosine alone would be extremely challenging.

8.3.1.5 Candida spp. and Amphotericin B

Detection of resistance to amphotericin B among *Candida* spp. using the CLSI M27-A3 broth microdilution method has been problematic due to the very narrow range of MIC values obtained [40, 42, 63]. In vitro and in vivo resistance clearly exists [64–72], and in vitro-in vivo correlations are possible [64, 66, 73, 74]. For example, Clancy and Nguyen examined data from a multicenter prospective study of candidemia and reported higher rates of therapeutic failure among patients who had isolates with amphotericin B MIC of ≥0.38 μg/mL (14/25 failed therapy (56%) versus only 12/74 (16%) with amphotericin B MICs of <0.38 μg/mL). In this study, Etest was used to determine the amphotericin B MIC. In general, agar-based methods such as Etest (AB BIODISK, Solna, Sweden) have proven to be the most sensitive and reliable means by which to detect resistance to amphotericin B among *Candida* species [19, 35, 63, 64, 66, 72, 73], although time-kill studies and determination of the minimum fungicidal concentration (MFC) may also be useful [68, 75]. Most recently, Park et al. examined *Candida* isolates from 107 candidemic patients treated with amphotericin B, and despite using five different methods of MIC determination, including Etest, they did not find any correlation of MIC with therapeutic outcome [63].

Despite the inconsistent data regarding the clinical relevance of the amphotericin B MIC, and although interpretive breakpoints have not been established, it is recognized that *Candida* for which MICs are >1 µg/mL are possibly "resistant" or may require higher doses of amphotericin B for optimal treatment [40, 53, 76]. Using the Etest agar-based technology as part of multicenter surveillance, we have identified differences in the susceptibility of the various species of *Candida* to amphotericin B [12, 13, 16, 19, 77–80]. Clearly, both *C. glabrata* and *C. krusei* exhibit decreased susceptibility to amphotericin B compared to that of *C. albicans* [12, 19, 66, 79, 80]. Furthermore, amphotericin B exhibits markedly delayed killing kinetics against these two species compared with that against *C. albicans* [75]. These findings are reflected in the treatment guidelines for *Candida* infections where higher doses of amphotericin B (≥0.7 mg/kg/day for *C. glabrata* and 1 mg/kg/day for *C. krusei*) are recommended for these two species [53, 76].

These issues are exemplified in a report by Krogh-Madsen and colleagues [66] who described a series of consecutive isolates of *C. glabrata* with increasing resistance to both amphotericin B and caspofungin recovered from a critically ill patient in an ICU. Amphotericin B MICs determined by Etest ranged from 1.5 to 32 µg/mL. An animal model documented therapeutic resistance to amphotericin B for isolates for which the amphotericin B MIC was 6–8 µg/mL (decreased response) and ≥32 µg/mL (fully resistant). The Etest method was superior to broth microdilution methods in predicting amphotericin B resistance. Reduced susceptibility to amphotericin B was also demonstrated by time-kill studies. This case not only demonstrates the development of amphotericin B resistance in *C. glabrata* but also confirms the Etest as a superior method for detecting such resistance.

Most other species of *Candida* remain susceptible to amphotericin B [12, 63]. Although notorious for developing clinical resistance to amphotericin B [73, 81–84], *C. lusitaniae* generally appears susceptible to this agent upon initial isolation from blood [12]. Thus, resistance to amphotericin B is not necessarily innate in this species but develops secondarily during treatment. *C. lusitaniae* has been shown to exhibit high-frequency phenotypic switching from amphotericin B susceptibility to resistance on exposure to the drug [85, 86]. A recent case report demonstrated the coexistence of two distinct color variants of *C. lusitaniae* upon subculture of positive blood samples onto CHROMagar Candida (Hardy Diagnostics, Santa Maria, CA) [73]. One variant (blue colony type) was susceptible to amphotericin B, and one (purple colony type) was resistant. The resistant variant was only detected after intense exposure to amphotericin B, therapy which was clinically unsuccessful. These results emphasize the importance of repeat amphotericin B susceptibility testing for patients with persistent *C. lusitaniae* infection [73].

8.3.1.6   Candida spp. and Echinocandins (Caspofungin, Micafungin, Anidulafungin)

The optimization of in vitro susceptibility testing of the echinocandins against *Candida* spp. has been difficult [17, 87, 88]; however, collaborative studies conducted by the CLSI Antifungal Subcommittee demonstrated that the use of RPMI 1640 broth medium, incubation at 35°C for 24 h, and an MIC endpoint criterion of prominent reduction in growth (MIC −2 or ≥50% inhibition relative to control growth) provided reproducible MIC results with separation of the "wild-type" MIC distribution from isolates with mutations in the *FKS*1gene for which reduced susceptibility was documented [17, 26, 88, 89].

The multicenter surveys conducted by Pfaller et al. [23, 26, 29, 90] and by Ostrosky-Zeichner et al. [58] both document the excellent potency and spectrum of all three echinocandins against more than 8,000 bloodstream infection isolates of *Candida* spp. Both surveys show that the common species *C. albicans, C. glabrata, C. tropicalis,* and *C. krusei* are highly susceptible to all three agents whereas elevated MICs (1 to 4 μg/mL) are seen for *C. parapsilosis* and *C. guilliermondii.* The available clinical data indicate that all of these species respond similarly to treatment with each of these agents [76, 91–100], although persistent fungemia with *C. parapsilosis* has been noted in the face of caspofungin therapy [97].

The CLSI applied the blueprint used to develop interpretive breakpoints for the azoles to develop breakpoints for anidulafungin, caspofungin, and micafungin when testing *Candida* spp. [34]. The CLSI Subcommittee considered (1) PK data documenting total serum drug concentrations above 1 μg/mL throughout the dosing interval for all three echinocandins [101]; (2) MIC distribution data showing that the MICs for >99% of 8,271 isolates was ≤2 μg/mL for all three agents [90]; (3) cross-resistance among the three echinocandins [90, 102] and lack thereof between echinocandins and azoles [17, 23, 26]; and (4) echinocandin MICs for *Candida* strains with documented *FKS*1 (glucan synthase) gene mutations and isolates from early case reports of echinocandin failures that were generally between 4 and 8 μg/mL or greater [102, 103]. Pharmacodynamic considerations included the fact that the indices associated with treatment efficacy for the echinocandins were AUC/MIC and Cmax/MIC ratios [104–106].

The clinical trial data (patient outcomes and baseline isolates) used in the initial CLSI analysis included four phase 2 or phase 3 studies of esophageal candidiasis treated with caspofungin [95], two phase 3 studies of invasive candidiasis treated with caspofungin [95], one phase 3 study of invasive candidiasis treated with anidulafungin [99], and two phase 3 studies of invasive candidiasis treated with micafungin [100, 107]. Clinical outcomes, as determined by the investigators at the end of therapy, were compared with the relevant echinocandin MICs for each baseline *Candida* isolate. Analysis of the available data supported an MIC of ≤2 μg/mL as predictive of efficacy (S) for all three echinocandins [34]. Successful outcomes at the breakpoint of ≤2 μg/mL were 88% for anidulafungin, 79% for caspofungin, and 80% for micafungin. An MIC predictive of resistance could not be defined for these agents based on data from clinical trials because of the paucity of isolates for which the MIC was greater than 2 µg/mL. The CLSI Subcommittee decided to recommend a "susceptible only" breakpoint of ≤2 µg/mL due to the lack of echinocandin resistance in the population of *Candida* isolates thus far [34]. Isolates of any species for which MICs are greater than 2 µg/mL were designated "nonsusceptible" (NS) and referred to a qualified reference laboratory for confirmation of identification and susceptibility to the echinocandins.

However, the first recommended CLSI clinical breakpoint for echinocandins and *Candida* spp. (≤2 µg/mL) is 8–64-fold higher than the "epidemiological cutoff values" (ECVs) for the echinocandins and most *Candida* spp., including *C. albicans*, *C. glabrata*, *C. tropicalis*, and *C. krusei* [90]. Only for *C. lusitaniae*, *C. parapsilosis*, and *C. guilliermondii* are the echinocandin ECVs within one or two dilutions of the clinical breakpoint [90]. This is important because while clinical breakpoints are designed to indicate those organisms most likely to respond to treatment with a given antimicrobial, ECVs are used as a sensitive measure to detect organisms that fall outside of the "wild-type" MIC distribution and may therefore exhibit resistance mechanisms. Although clinical trials showed that each of the three echinocandins could be used to treat invasive candidiasis due to *Candida* spp. isolates for which MICs were as high as the clinical breakpoint of 2 µg/mL [76, 91–100], several reports of clinical resistance to caspofungin therapy ([66, 103, 108–113]; Table 8.7), combined with studies of glucan synthase enzyme kinetics [113–115], suggested that the clinical breakpoint of ≤2 µg/mL needed to be adjusted to both predict clinical resistance as well as the emergence of strains with *FKS1* mutations. In each of the cases shown in Table 8.7, clinical failure of caspofungin therapy was associated with *FKS1* resistance mutations and MICs for all the echinocandins that were above the ECVs (e.g., elevated compared with "wild type") but not necessarily higher than the clinical breakpoint.

Therefore, the CLSI revisited the echinocandin breakpoints and proposed species-specific breakpoints that for several species (*C. albicans*, *C. glabrata*, *C. tropicalis*, and *C. krusei*) are significantly lower than previous breakpoints [116] (Table 8.8). These breakpoints will be more sensitive to detect emerging resistance to the echinocandins among common *Candida* spp. and should be better predictors of clinical failure [116].

8.3.2   Aspergillus spp.

There are currently no established AFST breakpoints for the *Aspergillus* spp. and any available antifungal agent. Animal (murine) model data have produced mixed results regarding the correlation of in vivo outcome from *Aspergillus* infection and in vitro AFST results for itraconazole [117, 118] and amphotericin B [119]. The most widely cited study supporting the clinical relevance of AFST results for invasive aspergillosis (IA) was published by Lass-Flörl et al. [120], who examined the relationship between MIC (determined according to methods similar to the CLSI M38-A) and outcome among 29 patients with IA. All six subjects with an amphotericin B MIC

Table 8.7 Clinical and in vitro resistance to echinocandins in patients with candidiasis[a]

Species (Ref.)	Infection type	Antifungal treatment[b]	MIC (μg/mL)[b]	Comments
C. glabrata [112]	Candidemia	CSF	CSF (2), ANF (0.5), MCF (0.25)	Mutation in FKS2, F659V
C. albicans [103]	Esophagitis	FLC, VRC, CSF, AMB	CSF (2), MCF (1)	Mutation in FKS1, F641S
C. tropicalis [113]	Candidemia	CSF, VRC	CSF (4), ANF (2), MCF (2)	Mutation, 50× increase in IC_{50}
C. tropicalis [113]	Candidemia	CSF, AMB	CSF (4), ANF (1), MCF (2)	Mutation, 50× increase in IC_{50}
C. tropicalis [113]	Candidemia	CSF, FLC	CSF (1), ANF (0.5), MCF (0.5)	Mutation, 38× increase in IC_{50}
C. albicans [111]	Esophagitis	CSF, AMB, FLC, VRC, ITZ, MCF	CSF (2), ANF (1), MCF (2)	Mutation, S645F and R1361H

[a]Adapted from Ref. [90]
[b]Abbreviations: AMB amphotericin B, ANF anidulafungin, CSF caspofungin, FLC fluconazole, ITZ itraconazole, MCF micafungin, VRC voriconazole, IC_{50} concentration that inhibits 50% of enzyme activity

Table 8.8 MIC interpretive breakpoints (BP) for the echinocandins and *Candida* species (CLSI)

Antifungal agent	Species	MIC BP (μg/mL)		
		S	I	R
Anidulafungin	*C. albicans*	≤0.25	0.5	≥1
	C. glabrata	≤0.25	0.5	≥1
	C. tropicalis	≤0.25	0.5	≥1
	C. krusei	≤0.25	0.5	≥1
	C. parapsilosis	≤2	4	≥8
	C. guilliermondii	≤2	4	≥8
Caspofungin	*C. albicans*	≤0.25	0.5	≥1
	C. glabrata	≤0.25	0.5	≥1
	C. tropicalis	≤0.25	0.5	≥1
	C. krusei	≤0.25	0.5	≥1
	C. parapsilosis	≤2	4	≥8
	C. guilliermondii	≤2	4	≥8
Micafungin	*C. albicans*	≤0.25	0.5	≥1
	C. glabrata	≤0.06	0.12	≥0.25
	C. tropicalis	≤0.25	0.5	≥1
	C. krusei	≤0.25	0.5	≥1
	C. parapsilosis	≤2	4	≥8
	C. guilliermondii	≤2	4	≥8

of ≤1 μg/mL survived their infection, while 22/23 with amphotericin B MIC of ≥2 μg/mL died [120]. Nine of the 22 who died were infected with *A. terreus*, a species well known to exhibit in vitro and in vivo resistance to amphotericin B [121, 122] (Table 8.9). Unfortunately, these results have not been replicated by other groups [123].

While there are insufficient data to establish clinical breakpoints for *Aspergillus* spp. based upon associations of MIC with clinical outcomes, there are several reports of treatment failure associated with elevated azole MICs [124–126]. Both primary resistance and secondary resistance have been demonstrated. Furthermore, studies of MIC distributions and resistance mechanisms in *A. fumigatus* have shown that MIC ECVs of ≤1 μg/mL for itraconazole and voriconazole and ≤0.25 μg/mL for posaconazole identified "wild-type" strains and provide separation of the wild-type population from those strains with resistance mutations of the cyp 51A gene [31, 127].

8.3.3   Other Moulds

Even fewer data exist to support clinical relevance of AFST results for moulds other than *Aspergillus* spp. The most useful information currently is the observation that

Table 8.9 List of fungal organism and antifungal combinations for which intrinsic or acquired resistance is commonly seen

Fungus	Drug(s)	Type of resistance
Aspergillus terreus	Amphotericin B	Intrinsic
Candida glabrata	Azoles	Intrinsic and acquired
	Amphotericin B	Acquired
Candida krusei	Fluconazole	Intrinsic
	Flucytosine	Intrinsic
	Amphotericin B	Acquired
Candida lusitaniae	Amphotericin B	Intrinsic and acquired
Histoplasma capsulatum	Fluconazole	Acquired
Scedosporium apiospermum	Amphotericin B	Intrinsic
Scedosporium prolificans	Amphotericin B	Intrinsic
Trichosporon spp.	Amphotericin B	Intrinsic

for certain individual organism and drug combinations, there is some consistency between the results obtained from in vitro testing and anecdotal or case series reports of clinical success. For example, posaconazole, the azole with the best in vitro activity against the zygomycetes [128], is also the azole that holds the greatest promise clinically in patients with zygomycosis [129, 130]. Conversely, there are organisms which exhibit in vitro resistance almost uniformly to agents that may have some clinical utility (e.g., *Fusarium* and voriconazole).

8.3.4   Cryptococcus neoformans

Although no AFST breakpoints have been established for this organism, there are data suggesting a relationship between fluconazole MIC and outcome of cryptococcal meningitis treated with this drug [131–133]. Aller et al. reported data from 25 patients with cryptococcal infection, noting that all five subjects who experienced clinical failure had fluconazole MICs of ≥16 μg/mL, while the other 20 subjects had fluconazole MICs of ≤8 μg/mL [133]. Similar data are not available for *C. neoformans* and other antifungal agents such as amphotericin B or flucytosine.

A more recently published analysis of 74 cases of cryptococcal meningitis failed to find any association between MIC and clinical outcome of treatment with amphotericin B, flucytosine, or fluconazole [134]. However, this study is limited by the inclusion of only 15 persons treated with fluconazole.

8.3.5   Other Yeast, Yeast-Like Fungi, and Dimorphic Fungi

There are insufficient data for the less common yeast and yeast-like fungi to support clinical utility of AFST results. Among the dimorphic fungi, there is one report suggesting an association between fluconazole MIC and clinical response to fluconazole therapy among HIV-infected subjects with disseminated histoplasmosis [135].

8.4   A Practical Approach to the Use of Antifungal Susceptibility Testing

Given the data we have reviewed in this chapter, how then should one use AFST results in the care of patients? Table 8.10 provides detailed guidance on the use of antifungal susceptibility testing in the clinical laboratory.

As additional data are gathered and relevant breakpoints established for new organism-drug combinations (e.g., *Aspergillus* spp. and azoles), the recommendations in Table 8.10 are subject to modification.

8.5   Conclusion

Standardized methods for antifungal susceptibility testing have been developed and are now widely available. Do the results of such testing have any utility in the care of patients with invasive fungal infections? In this chapter, we review the available published data addressing the clinical relevance of antifungal susceptibility test results. By far the most data exist to support the clinical relevance of AFST results for *Candida* against fluconazole, and these data suggest that the clinical utility of this information mirrors that put forward for antibacterial susceptibility testing. Clinical relevance has also been demonstrated for selected other antifungal agents against *Candida* and *Cryptococcus* spp. By contrast, little direct support for the clinical utility of AFST for moulds is available.

Since antifungal susceptibility patterns are strongly associated with species identification, we recommend that all laboratories identify invasive fungal isolates to the species level. In addition, laboratories should routinely test invasive (sterile site) *Candida glabrata* isolates against fluconazole and an echinocandin. Additional testing of other fungal organisms and other drugs are also recommended in some situations, as dictated by the infection site, the organism involved, and the clinical response to therapy. These recommendations will evolve as additional data is gathered and relevant breakpoints established for new organism-drug combinations. We anticipate that AFST will play an increasingly important role in the management of patients with invasive fungal infections.

Table 8.10 Recommendations for use of antifungal susceptibility testing in the clinical laboratory[a]

Clinical setting	Recommendation
Routine	• Species level identification of all *Candida* isolates from deep sites (e.g., blood, normally sterile fluids, tissues, abscesses)
	• Species level identification of *Aspergillus*, genus level for all other moulds
	• Species level identification of noncandidal yeast isolates from deep sites
	• Routine antifungal testing of fluconazole and an echinocandin against *C. glabrata* from deep sites
	• Routine testing of fluconazole and an echinocandin against other species of *Candida* may be helpful but susceptibility usually predictable
Oropharyngeal candidiasis	• Determination of azole susceptibility not routinely necessary
	• Susceptibility testing may be useful for patients unresponsive to therapy
Invasive disease with clinical failure of initial therapy	• Consider susceptibility testing as an adjunct
	– *Candida* species and amphotericin B, flucytosine, fluconazole, voriconazole, and an echinocandin
	– *C. neoformans* and fluconazole, flucytosine, amphotericin B
	– *Aspergillus* species and amphotericin B, posaconazole, itraconazole, voriconazole
	• Consultation with an experienced microbiologist recommended

Infection with species with high rates of intrinsic or acquired resistance

- Susceptibility testing not necessary when intrinsic resistance is known
  - *A. terreus* and amphotericin B
  - *C. krusei* and fluconazole, flucytosine
  - *Cryptococcus*, *Rhodotorula*, and *Trichosporon* and echinocandins
  - *Rhodotorula* and azoles
  - *Scedosporium apiospermum* and amphotericin B
  - *Scedosporium prolificans* and amphotericin B, azoles, echinocandins
  - *Zygomycetes* and voriconazole, echinocandins
- When high rates of acquired resistance, monitor closely for signs of failure and perform susceptibility testing
  - *C. glabrata* and fluconazole, amphotericin B
  - *C. krusei* and amphotericin B
  - *C. lusitaniae* and amphotericin B
  - *C. rugosa* and amphotericin B

New treatment options (e.g., echinocandins, voriconazole, posaconazole) or unusual organisms

- Susceptibility of *Candida* spp. to echinocandins may be assumed unless initial response is suboptimal
- Susceptibility testing of voriconazole and posaconazole may be helpful (especially versus *Aspergillus* spp.)
- Selection of therapy based on published consensus guidelines [135–141] and review of survey data on the organism-drug combination in question
- Susceptibility testing may be helpful when patient is not responding to what should be effective therapy

(continued)

Table 8.10 (continued)

Clinical setting	Recommendation
Patients who respond to therapy despite being infected with an organism later found to be resistant	• Best approach not clear • Take into account severity of infection, patient immune status, consequences of recurrent infection, etc. • Consider alternative therapy for infections with isolates that appear to be highly resistant to therapy selected
Mould infections	• Identification to genus and species desirable • Routine susceptibility testing not recommended • Susceptibility of *Aspergillus* to itraconazole, posaconazole, and voriconazole may help sort out cross-resistance • Resistance to amphotericin B may help identify *Trichosporon asahii* • Clinical interpretive criteria have not been established for any agents

Selection of susceptibility testing methods	• Standardized methods
	• CLSI broth-based methods
	  – Yeasts; M27-A3
	  – Moulds; M38-A2
	• CLSI agar-based methods
	  – Disk diffusion; yeasts M44-A2
	  – Disk diffusion; moulds M51-A
	• Commercial methods
	  – ATB FUNGUS 2
	  – Etest
	  – Sensititre YeastOne
	  – VITEK 2

[a]Adapted from Refs. [40] and [53]

References

1. Clinical and Laboratory Standards Institute (formerly National Committee for Clinical Laboratory Standards) (2008) Reference method for broth dilution antifungal susceptibility testing of yeasts: approved standard, 3rd edn. M27-A3. Clinical and Laboratory Standards Institute, Wayne

2. Rodriguez-Tudela JL, Barchiesi F, Bille J et al (2003) Method for the determination of minimum inhibitory concentration by broth dilution of fermentative yeasts. Clin Microbiol Infect 9:1–8

3. Clinical and Laboratory Standards Institute (formerly National Committee for Clinical Laboratory Standards) (2008) Reference method for broth dilution antifungal susceptibility testing of filamentous fungi. Approved Standard 2nd edn. M38-A2. Clinical and Laboratory Standards Institute, Wayne

4. Rodriguez-Tudela JL, Donnelly JP, Arendrup MC et al (2008) EUCAST (European Committee for Antimicrobial Susceptibility Testing) technical note on the method for the determination of broth dilution minimum inhibitory concentrations of antifungal agents for conidia-forming moulds. Clin Microbiol Infect 14:982–984

5. Clinical and Laboratory Standards Institute (formerly National Committee for Clinical Laboratory Standards) (2004) Methods for antifungal disk diffusion susceptibility testing of yeasts: approved guideline, M44-A. Clinical and Laboratory Standards Institute, Wayne

6. Clinical and Laboratory Standards Institute (formerly National Committee for Clinical Laboratory Standards) (2008) Method for antifungal disk diffusion susceptibility testing of filamentous fungi: proposed guideline. M51-P. Clinical and Laboratory Standards Institute, Wayne

7. Barry AL, Pfaller MA, Brown SD et al (2000) Quality control limits for broth microdilution susceptibility tests of ten antifungal agents. J Clin Microbiol 38:3457–3459

8. Barry AL, Bille J, Brown S et al (2003) Quality control limits for fluconazole disk susceptibility tests on Mueller-Hinton agar with glucose and methylene blue. J Clin Microbiol 41:3410–3412

9. Krisher K, Brown SD, Traczewski MM (2004) Quality control parameters for broth microdilution tests of anidulafungin. J Clin Microbiol 42:490

10. Pfaller MA, Diekema DJ (2002) Role of sentinel surveillance of candidemia: trends in species distribution and antifungal susceptibility. J Clin Microbiol 40:3551–3557

11. Pfaller MA, Messer SA, Boyken L et al (2002) In vitro activities of 5-fluorocytosine against 8,803 clinical isolates of *Candida* spp.: global assessment of primary resistance using National Committee for Clinical Laboratory Standards susceptibility testing methods. Antimicrob Agents Chemother 46:3518–3521

12. Pfaller MA, Messer SA, Hollis RJ et al (2002) In vitro activities of ravuconazole and voriconazole compared with those of four approved systemic antifungal agents against 6,970 clinical isolates of *Candida* spp. Antimicrob Agents Chemother 46:1723–1727

13. Pfaller MA, Diekema DJ, Jones RN, The SENTRY Participants Group et al (2002) Trends in antifungal susceptibility of *Candida* spp. isolated from pediatric and adult patients with bloodstream infections: SENTRY Antimicrobial Surveillance Program, 1997 to 2000. J Clin Microbiol 40:852–856

14. Pfaller MA, Messer SA, Boyken L et al (2003) Variation in susceptibility of bloodstream isolates of *Candida glabrata* to fluconazole according to patient age and geographic location. J Clin Microbiol 41:2176–2179

15. Pfaller MA, Diekema DJ (2004) Rare and emerging opportunistic fungal pathogens: concern for resistance beyond *Candida albicans* and *Aspergillus fumigatus*. J Clin Microbiol 42:4419–4431

16. Pfaller MA, Diekema DJ (2004) Twelve years of fluconazole in clinical practice: global trends in species distribution and fluconazole susceptibility of bloodstream isolates of *Candida*. Clin Microbiol Infect 10(suppl 1):11–23

17. Pfaller MA, Messer SA, Boyken L et al (2004) Further standardization of broth microdilution methodology for in vitro susceptibility testing of caspofungin against *Candida* species by sue of an international collection of more than 3,000 clinical isolates. J Clin Microbiol 42:3117–3119

18. Pfaller MA, Hazen KC, Messer SA et al (2004) Comparison of results of fluconazole disk diffusion testing for *Candida* species with results from a central reference laboratory in the ARTEMIS Global Antifungal Surveillance Program. J Clin Microbiol 42:3607–3612

19. Pfaller MA, Boyken L, Messer SA et al (2004) Evaluation of the Etest method using Mueller-Hinton agar with glucose and methylene blue for determining amphotericin B MICs for 4,936 clinical isolates of *Candida* species. J Clin Microbiol 42:4977–4979

20. Pfaller MA, Messer SA, Boyken L et al (2004) Geographic variation in the susceptibilities of invasive isolates of *Candida glabrata* to seven systemically active antifungal agents: a global assessment from the ARTEMIS Antifungal Surveillance Program conducted in 2001 and 2002. J Clin Microbiol 42:3142–3146

21. Pfaller MA, Messer SA, Boyken L, Hollis RJ et al (2004) In vitro activities of voriconazole, posaconazole, and fluconazole against 4,169 clinical isolates of *Candida* spp. and *Cryptococcus neoformans* collected during 2001 and 2002 in the ARTEMIS global antifungal surveillance program. Diagn Microbiol Infect Dis 48:201–205

22. Pfaller MA, Boyken L, Hollis RJ et al (2005) In vitro susceptibilities of clinical isolates of *Candida* species, *Cryptococcus neoformans*, and *Aspergillus* species to itraconazole: global survey of 9,359 isolates tested by Clinical and Laboratory Standards Institute broth microdilution methods. J Clin Microbiol 43:3807–3810

23. Pfaller MA, Boyken L, Hollis RJ et al (2005) In vitro activities of anidulafungin against more than 2,500 clinical isolates of *Candida* spp., including 315 isolates resistant to fluconazole. J Clin Microbiol 43:5425–5427

24. Pfaller MA, Diekema DJ, Rinaldi MG, The Global Antifungal Surveillance Group et al (2005) Results from the ARTEMIS DISK Global Antifungal Surveillance Study: a 6.5-year analysis of susceptibilities of *Candida* and other yeast species to fluconazole and voriconazole by standardized disk diffusion testing. J Clin Microbiol 43:5848–5859

25. Pfaller MA, Boyken L, Messer SA et al (2005) Comparison of results of voriconazole disk diffusion testing for *Candida* species with results from a central reference laboratory in the ARTEMIS Global Antifungal Surveillance Program. J Clin Microbiol 43:5208–5213

26. Pfaller MA, Boyken L, Hollis RJ et al (2006) In vitro susceptibilities of *Candida* spp. to caspofungin: four years of global surveillance. J Clin Microbiol 44:760–763

27. Pfaller MA, Diekema DJ, Rex JH et al (2006) Correlation of MIC with outcome for *Candida* species tested against voriconazole: analysis and proposal for interpretive breakpoints. J Clin Microbiol 44:819–826

28. Pfaller MA, Diekema DJ, Sheehan DJ (2006) Interpretive breakpoints for fluconazole and *Candida* revisited: a blueprint for the future of antifungal susceptibility testing. Clin Microbiol Rev 19:435–447

29. Pfaller MA, Boyken L, Hollis RJ et al (2006) Global surveillance of the in vitro activity of micafungin against *Candida*: a comparison with caspofungin using Clinical and Laboratory Standards Institute recommended methods. J Clin Microbiol 44:3533–3538

30. Pfaller MA, Messer SA, Hollis RJ, Boyken L, Tendolkar S, Kroeger J, Diekema DJ (2009) Variation in susceptibility of bloodstream isolates of Candida glabrata to fluconazole according to patient age and geographic location in the U.S., 2001–2007. J Clin Microbiol 47:3185–3190

31. Pfaller MA, Diekema DJ, Ghannoum A et al (2009) Wild type MIC distribution and epidemiologic cutoff values for *Aspergillus fumigatus* and three triazoles as determined by the CLSI broth microdilution methods. J Clin Microbiol 47:3142–3146

32. Diekema DJ, Messer SA, Boyken LD, Hollis RJ, Kroeger J, Tendolkar S, Pfaller MA (2009) In vitro activity of seven systemically active antifungal agents against a large global collection of rare Candida species as determined by CLSI broth microdilution methods. J Clin Microbiol 47:3170–3177

33. Diekema DJ, Messer SA, Hollis RJ, Boyken L, Tendolkar S, Kroeger J, Jones RN, Pfaller MA (2009) A global evaluation of voriconazole activity tested against recent clinical isolates of Candida spp. Diagn Microbiol Infect Dis 63:233–236

34. Pfaller MA, Diekema DJ, Ostrosky-Zeichner L et al (2008) Correlation of MIC with outcome for Candida species tested against caspofungin, anidulafungin, and micafungin: analysis and proposal for interpretive MIC breakpoints. J Clin Microbiol 46:2620–2629

35. Pfaller MA, Messer SA, Bolmstrom A (1998) Evaluation of Etest for determining in vitro susceptibility of yeast isolates to amphotericin B. Diagn Microbiol Infect Dis 32:223–227

36. Espinel Ingroff A, Pfaller MA, Messer SA et al (2004) Multicenter comparison of the Sensititre YeastOne colorimetric antifungal panel with the NCCLS M27-A2 reference method for testing new antifungal agents against clinical isolates of Candida spp. J Clin Microbiol 42:718–721

37. Pfaller MA, Diekema DJ, Procop GW, Rinaldi MG (2007) Multicenter comparison of the VITEK 2 yeast susceptibility test with the CLSI broth microdilution reference method for testing fluconazole against Candida spp. J Clin Microbiol 45:796–802

38. Pfaller MA, Diekema DJ, Procop GW, Rinaldi MG (2007) Multicenter comparison of the VITEK 2 yeast susceptibility test with the CLSI broth microdilution reference method for testing amphotericin B, flucytosine, and voriconazole against Candida spp. J Clin Microbiol 45:3522–3528

39. Torres-Rodriguez JM, Alvarado-Ramirez E (2007) In vitro susceptibilities to yeasts using the ATB FUNGUS 2 method compared to Sensititre YeastOne and standard CLSI M27-A2 methods. J Antimicrob Chemother 60:658

40. Rex JH, Pfaller MA (2002) Has antifungal susceptibility testing come of age? Clin Infect Dis 35:982–989

41. Rex JH, Pfaller MA, Galgiani JN et al (1997) Development of interpretive breakpoints for antifungal susceptibility testing: conceptual framework and analysis of in vitro – in vivo correlation data for fluconazole, itraconazole, and Candida infections. Clin Infect Dis 24:235–247

42. Rex JH, Pfaller MA, Walsh TJ et al (2001) Antifungal susceptibility testing: practical aspects and current challenges. Clin Microbiol Rev 14:643–658

43. Andes D (2003) Clinical pharmacodynamics of antifungals. Infect Dis Clin N Am 17:635–649

44. Andes D (2003) In vivo pharmacodynamics of antifungal drugs in treatment of candidiasis. Antimicrob Agents Chemother 47:1179–1186

45. Andes D, Marchill K, Lawther J et al (2003) In vivo pharmacodynamics of HMR 3270, a glucan synthase inhibitor, in a murine candidiasis model. Antimicrob Agents Chemother 47:1187–1192

46. Voss A, de Pauw BE (1999) High-dose fluconazole therapy in patients with severe fungal infections. Eur J Clin Microbiol Infect Dis 18:165–174

47. Morrell M, Fraser VJ, Kollef MJ (2005) Delaying empiric treatment of Candida bloodstream infection until positive blood culture results are obtained: a potential risk factor for mortality. Antimicrob Agents Chemother 49:3640

48. Garey KW, Rege M, Pai MP et al (2006) Time to initiation of fluconazole therapy impacts mortality in patients with candidemia: a multi-institutional study. Clin Infect Dis 43:25

49. Collins CD, Eschenauer GA, Salo SL, Newton DW (2007) To test or not to test: a cost minimization analysis of susceptibility testing for patients with documented Candida glabrata fungemias. J Clin Microbiol 45:1884

50. Perkins MD, Sabuda DM, Elsayed S, Laupland KB (2007) Adequacy of empirical antifungal therapy and effect of outcome among patients with invasive Candida species infections. J Antimicrob Chemother 60:613

51. Ostrosky-Zeichner L, Rex JH, Pfaller MA et al (2008) Rationale for reading fluconazole MICs at 24h rather than 48h when testing Candida spp. by the CLSI M27-A2 standard method. Antimicrob Agents Chemother 52:4175–4177

52. Pfaller MA, Boyken LB, Hollis RJ et al (2008) Validation of 24-hour fluconazole MIC readings versus the CLSI 48-hour broth microdilution reference method: results from a global Candida antifungal surveillance program. J Clin Microbiol 46:3585–3590

53. Pappas PG, Kauffman CA, Andes D et al (2009) Clinical practice guidelines for the management of candidiasis: 2009 update by the Infectious Diseases Society of America. Clin Infect Dis 48:503–535

54. Andes D, van Ogtrop M (1999) Characterization and quantitation of the pharmacodynamics of fluconazole in a neutropenic murine disseminated candidiasis infection model. Antimicrob Agents Chemother 43:2116–2120

55. Louie A, Drusano GL, Banerjee P et al (1998) Pharmacodynamics of fluconazole in a murine model of systemic candidiasis. Antimicrob Agents Chemother 42:1105–1109

56. Rodriguez-Tudela JL, Almirante B, Rodriguez-Pardo D et al (2007) Correlation of the MIC and Dose/MIC ratio of fluconazole to the therapeutic response of patients with mucosal candidiasis and candidemia. Antimicrob Agents Chemother 51:3599–3604

57. Cuesta I, Bielza C, Larranaga P et al (2009) Data mining validation of fluconazole breakpoints established by the European Committee on Antimicrobial Susceptibility Testing. Antimicrob Agents Chemother 53:2949–2954

58. Pfaller MA, Andes D, Diekema DJ, Espinel-Ingroff A, Sheehan D, The CLSI Subcommittee for Antifungal Susceptibility Testing (2010) Wild-type MIC distributions, epidemiological cutoff values and species-specific clinical breakpoints for fluconazole and *Candida*: time for harmonization of CLSI and EUCAST broth microdilution methods. Drug Resist Updat 13:180–195

59. Ostrosky-Zeichner L, Oude Lashof AML, Kullberg BJ (2003) Voriconazole salvage treatment of invasive candidiasis. Eur J Clin Microbiol Infect Dis 22:651–655

60. Kartsonis NA, Saah A, Lipka CJ et al (2004) Second-line therapy with caspofungin for mucosal or invasive candidiasis: results from the caspofungin compassionate use study. J Antimicrob Chemother 53:878–881

61. Rodriguez-Tudela JL, Donnelly JP, Arendrup MC et al (2008) EUCAST technical note on voriconazole. Clin Microbiol Infect 14:985–987

62. Pfaller MA, Andes D, Arendrup MC et al (2011) Clinical breakpoints for voriconazole and Candida spp. revisited: review of microbiologic, molecular, pharmacodynamic, and clinical data as they pertain to the development of species-specific interpretive criteria. Diagn Microbiol Infect Dis 70:330–343

63. Park BJ, Arthington-Skaggs BA, Hajjeh RA et al (2006) Evaluation of amphotericin B interpretive breakpoints for *Candida* bloodstream isolates by correlation with therapeutic outcome. Antimicrob Agents Chemother 50:1287–1292

64. Clancy CJ, Nguyen MH (1999) Correlation between in vitro susceptibility determined by Etest and response to therapy with amphotericin B: results from a multicenter prospective study of candidemia. Antimicrob Agents Chemother 43:1289–1290

65. Gibbs WJ, Drew RH, Perfect JR (2005) Liposomal amphotericin B: clinical experience and perspectives. Expert Rev Anti Infect Ther 3:167–181

66. Krogh-Madsen M, Arendrup MC, Heslet L et al (2006) Amphotericin B and caspofungin resistance in *Candida glabrata* isolates recovered from a critically ill patient. Clin Infect Dis 42:938–944

67. Law D, Moore CB, Denning DW (1997) Amphotericin B resistance testing of *Candida* spp.: a comparison of methods. J Antimicrob Chemother 40:109–112

68. Nguyen MH, Clancy CJ, Yu VL et al (1998) So in vitro susceptibility data predict the microbiologic response to amphotericin B? Results of a prospective study of patients with *Candida* fungemia. J Infect Dis 177:425–430

69. Nolte FS, Parkinson T, Falconer DJ et al (1997) Isolation and characterization of fluconazole- and amphotericin B-resistant *Candida albicans* from blood of two patients with leukemia. Antimicrob Agents Chemother 44:196–199

70. Sterling TR, Gasser RA, Ziegler A (1996) Emergence of resistance to amphotericin B during therapy for *Candida glabrata* infection in an immunocompetent host. Clin Infect Dis 23:187–188

71. Sterling T, Merz WG (1998) Resistance to amphotericin B: emerging clinical and microbiological patterns. Drug Resist Updat 1:161–165

72. Wanger A, Mills K, Nelson PW et al (1995) Comparison of Etest and National Committee for Clinical Laboratory Standards broth macrodilution method for antifungal susceptibility testing: enhanced ability to detect amphotericin B-resistant *Candida* isolates. Antimicrob Agents Chemother 39:2520–2522

73. McClenny NB, Fei H, Baron EJ et al (2002) Change in colony morphology of *Candida lusitaniae* in association with development of amphotericin B resistance. Antimicrob Agents Chemother 46:1325–1328

74. O'Day M, Ray WA, Robinson RD et al (1987) Correlation of in vitro and in vivo susceptibility of *Candida albicans* to amphotericin B and natamycin. Investig Ophthalmol Vis Sci 29:596–603

75. Canton E, Peman J, Gobernado M et al (2004) Patterns of amphotericin B killing kinetics against seven *Candida* species. Antimicrob Agents Chemother 48:2477–2482

76. Spellberg BJ, Filler SG, Edwards JE Jr (2006) Current treatment strategies for disseminated candidiasis. Clin Infect Dis 42:244–251

77. Blignant E, Molepo J, Pujol C et al (2005) Clade-related amphotericin B resistance among South African *Candida albicans* isolates. Diagn Microbiol Infect Dis 53:29–31

78. Hajjeh RA, Sofair AN, Harrison IH et al (2004) Incidence of bloodstream infections due to *Candida* species and in vitro susceptibilities of isolates collected from 1998 to 2000 in a population-based active surveillance program. J Clin Microbiol 42:1519–1527

79. Kao AS, Brandt ME, Pruitt WR et al (1999) The epidemiology of candidemia n two United States cities: results of a population-based active surveillance. Clin Infect Dis 29:1164–1170

80. Yang YL, Li SY, Chang HH (2005) Susceptibilities to amphotericin B and fluconazole of *Candida* species in TSARY 2002. Diagn Microbiol Infect Dis 51:179–183

81. Favel A, Michel-Nguyen A, Datry A et al (2004) Susceptibility of clinical isolates of C. lusitaniae to five systemic antifungal agents. J Antimicrob Chemother 53:526–529

82. Hawkins JL, Baddour LM (2003) *Candida lusitaniae* infections in the era of fluconazole availability. Clin Infect Dis 36:e14–e18

83. Minari A, Hachem R, Raad I (2001) *Candida lusitaniae*: a cause of breakthrough fungemia in cancer patients. Clin Infect Dis 32:186–190

84. Peyron F, Favel A, Michel-Nguyen A et al (2001) Improved detection of amphotericin B-resistant isolates of *Candida lusitaniae* by Etest. J Clin Microbiol 39:339–342

85. Miller NS, Dick JD, Merz WG (2006) Phenotypic switching in *Candida lusitaniae* on copper sulfate indicator agar: association with amphotericin B resistance and filamentation. J Clin Microbiol 44:1536–1539

86. Yoon SA, Vazquez JA, Stefan PE et al (1999) High-frequency, in vitro reversible switching of *Candida lusitaniae* clinical isolates from amphotericin B susceptibility to resistance. Antimicrob Agents Chemother 43:836–845

87. Bartizal K, Odds FC (2003) Influences of methodological variables on susceptibility testing of caspofungin against *Candida* species and *Aspergillus fumigatus*. Antimicrob Agents Chemother 47:2100–2107

88. Odds FC, Motyl M, Andrade R et al (2004) Interlaboratory comparison of results of susceptibility testing with caspofungin against *Candida* and *Aspergillus* species. J Clin Microbiol 42:3475–3482

89. Park S, Kelly R, Kahn JN et al (2005) Specific substitutions in the echinocandins target Fks1p account for reduced susceptibility of rare laboratory and clinical *Candida* sp. isolates. Antimicrob Agents Chemother 49:3264–3273

90. Pfaller MA, Boyken L, Hollis RJ, Kroeger J, Messer S, Tendolkar S, Jones RN, Turnidge J, Diekema DJ (2010) Wild-type MIC distributions and epidemiological cutoff values (ECVs) for the echinocandins and *Candida* spp. J Clin Microbiol 48(1):52–56

91. Chandrasekar PH, Sobel JD (2006) Micafungin: a new echinocandins. Clin Infect Dis 42:1171–1178

92. Colombo AL, Melo ASA, Rosas RFC et al (2003) Outbreak of *Candida rugosa* candidemia: an emerging pathogen that may be refractory to amphotericin B therapy. Diagn Microbiol Infect Dis 46:253–257

93. Colombo AL, Perect J, DiNubile M et al (2003) Global distribution and outcomes for *Candida* species causing invasive candidiasis: results from an international randomized double-blind study of caspofungin versus amphotericin B for the treatment of invasive candidiasis. Eur J Clin Microbiol Infect Dis 22:470–474
94. Glasmacher A, Cornely OA, Orlopps K et al (2006) Caspofungin treatment in severely ill, immunocompromised patients: a case-documentation study of 118 patients. J Antimicrob Chemother 57:127–134
95. Kartsonis N, Killar J, Mixson L et al (2005) Caspofungin susceptibility testing of isolates from patients with esophageal candidiasis or invasive candidiasis: relationship of MIC to treatment outcome. Antimicrob Agents Chemother 49:3616–3623
96. Krause DS, Reinhardt J, Vazquez JA et al (2004) Phase, randomized dose-ranging study evaluating the safety and efficacy of anidulafungin in invasive candidiasis and candidemia. Antimicrob Agents Chemother 48:2021–2024
97. Mora-Duarte J, Betts R, Rotstein C et al (2002) Comparison of caspofungin and amphotericin B for invasive candidiasis. N Engl J Med 347:2020–2029
98. Ostrosky-Zeichner L, Kontoyiannis D, Raffalii J et al (2005) International, open-label, noncomparative, clinical trial of micafungin alone and in combination for treatment of newly diagnosed and refractory candidemia. Eur J Clin Microbiol Infect Dis 24:654–661
99. Reboli AC, Rotstein C, Pappas PG et al (2007) Anidulafungin versus fluconazole for invasive candidiasis. N Engl J Med 356:2472–2482
100. Pappas PG, Rotstein CM, Betts RF et al (2007) Micafungin versus caspofungin for treatment of candidemia and other forms of invasive candidiasis. Clin Infect Dis 45:883–893
101. Cappelletty D, Eiselstein-McKitrick K (2007) The echinocandins. Pharmacotherapy 27(3):369
102. Perlin DS (2007) Resistance to echinocandins-class antifungal drugs. Drug Resist Updat 10:121
103. Baixench MT, Aoun N, Desnos-Ollivier M et al (2007) Acquired resistance to echinocandins in *Candida albicans*: case report and review. J Antimicrob Chemother 59:1076
104. Louie A, Deziel M, Liu W et al (2005) Pharmacodynamics of caspofungin in a murine model of systemic candidiasis: importance of persistence of caspofungin in tissues to understanding drug activity. Antimicrob Agents Chemother 49:5058–5068
105. Andes D, Diekema DJ, Pfaller MA et al (2008) In vivo pharmacodynamic characterization of anidulafungin in a neutropenic murine candidiasis model. Antimicrob Agents Chemother 52:539–550
106. Andes D, Diekema DJ, Pfaller MA et al (2008) In vivo pharmacodynamic target investigation for micafungin against *C. albicans* and *C. glabrata* in a neutropenic murine candidiasis model. Antimicrob Agents Chemother 52:3497–3503
107. Kuse ER, Chutchotisakd P, da Cunha CA et al (2007) Micafungin versus liposomal amphotericin B for candidemia and invasive candidiasis: a phase III randomized double-blind trial. Lancet 369:1519
108. Hernandez S, Lopez-Ribot JL, Najvor LK et al (2004) Caspofungin resistance in *Candida albicans*: correlating clinical outcome with laboratory susceptibility testing of three isogenic isolates serially obtained from a patient with progressive *Candida* esophagitis. Antimicrob Agents Chemother 48:1382–1383
109. Dodgson KJ, Dodgson AR, Pujol C et al (2005) Caspofungin resistant *C. glabrata*. Clin Microbiol Infect 11(suppl 2):364
110. Moudgal V, Little T, Boikov D et al (2005) Multiechinocandin- and multiazole-resistant *Candida parapsilosis* isolates serially obtained during therapy for prosthetic valve endocarditis. Antimicrob Agents Chemother 49:767–769
111. Laverdiere M, Lalonde RG, Baril JG et al (2006) Progressive loss of echinocandins activity following prolonged use for treatment of *Candida albicans* oesophagitis. J Antimicrob Chemother 57:705–708
112. Thompson GR, Wiederhold NP, Vallor AC et al (2008) Development of caspofungin resistance following prolonged therapy for invasive candidiasis secondary to Candida glabrata infection. Antimicrob Agents Chemother 52:3783–3785

113. Garcia-Effron G, Kontoyiannis DP, Lewis RE, Perlin DS (2008) Caspofungin-resistant *Candida tropicalis* strains causing breakthrough fungemia in patients at high risk for hematologic malignancies. Antimicrob Agents Chemother 52:4181–4183

114. Garcia-Effron G, Park S, Perlin DS (2009) Correlating echinocandin MIC and kinetic inhibition of FKS1 mutant glucan synthases for Candida albicans: implications for interpretive breakpoints. Antimicrob Agents Chemother 53:112–122

115. Garcia-Effron G, Lee S, Park S et al (2009) Effect of Candida glabrata FKS1 and FKS2 mutations on echinocandin sensitivity and kinetics of 1, 3-beta-D glucan synthase: implication for the existing susceptibility breakpoint. Antimicrob Agents Chemother 53:3690–3699

116. Pfaller MA, Diekema DJ, Andes D, Arendrup MC, Brown SD, Lockhart SR, Motyl M, Perlin D, The CLSI Subcommittee for Antifungal Testing (2011) Clinical breakpoints for the echinocandins and Candida revisited: integration of molecular, clinical, and microbiological data to arrive at species-specific interpretive criteria. Drug Resist Updat 14:164–176

117. Denning DW, Radford SA, Oakley KL et al (1997) Correlation between in-vitro susceptibility testing to itraconazole and in-vivo outcome of *Aspergillus fumigatus* infection. J Antimicrob Chemother 40:401–414

118. Mosquera J, Warn PA, Morrissey J et al (2001) Susceptibility testing of *Aspergillus flavus*: inoculum dependence with itraconazole and lack of correlation between susceptibility to amphotericin B in vitro and outcome in vivo. Antimicrob Agents Chemother 45:1456–1462

119. Johnson EM, Oakley KL, Radford SA et al (2000) Lack of correlation of in vitro amphotericin B susceptibility testing with outcome in a murine model of *Aspergillus* infection. J Antimicrob Chemother 45:85–93

120. Lass-Florl C, Kofler G, Kropshofer G et al (1998) In-vitro testing of susceptibility to amphotericin B is a reliable predictor of clinical outcome in invasive aspergillosis. J Antimicrob Chemother 42:497–502

121. Steinbach WJ, Benjamin DK, Kontoyiannis DP et al (2004) Infections due to *Aspergillus terreus*: a multicenter retrospective analysis of 83 cases. Clin Infect Dis 39:192–198

122. Steinbach WJ, Perfect JR, Schell WA et al (2004) In vitro analyses, animal models, and 60 clinical cases of invasive *Aspergillus terreus* infection. Antimicrob Agents Chemother 48:3217–3225

123. Lionakis MS, Lewis RE, Chamilos G et al (2005) *Aspergillus* susceptibility testing in patients with cancer and invasive aspergillosis: difficulties in establishing correlation between in vitro susceptibility data and the outcome of initial amphotericin B therapy. Pharmacotherapy 25:1174–1180

124. Howard SJ, Cerar D, Anderson MJ et al (2009) Frequency and evolution of azole resistance in Aspergillus fumigatus associated with treatment failure. Emerg Infect Dis 15:1068–1076

125. Denning DW, Vankateswarlu K, Oakley KL et al (1997) Itraconazole resistance in Aspergillus fumigatus. Antimicrob Agents Chemother 41:1364–1368

126. Verweij PE, Mellado E, Melchers WJG (2007) Multiple-triazole-resistant aspergillosis. N Engl J Med 356:1481

127. Rodriguez-Tudela JL, Alcazar-Fuoli L, Mellado E et al (2008) Epidemiological cutoffs and cross-resistance to azole drugs in *Aspergillus fumigatus*. Antimicrob Agents Chemother 52:2468

128. Diekema DJ, Messer SA, Hollis RJ et al (2003) Activities of caspofungin, itraconazole, posaconazole, ravuconazole, voriconazole, and amphotericin B against 448 recent clinical isolates of filamentous fungi. J Clin Microbiol 41:3623–3626

129. Greenberg RN, Mullane K, van Burik JA et al (2006) Posaconazole as salvage therapy for zygomycosis. Antimicrob Agents Chemother 50:126–133

130. van Burik JAH, Hare RS, Solomon HF et al (2006) Posaconazole is effective as salvage therapy in zygomycosis: a retrospective summary of 91 cases. Clin Infect Dis 42:e61–e65

131. Jessup CJ, Pfaller MA, Messer SA et al (1998) Fluconazole susceptibility testing of Cryptococcus neoformans: comparison of two broth microdilution methods and clinical correlates among isolates from Ugandan AIDS patients. J Clin Microbiol 36:2874–2876

132. Witt MD, Lewis RJ, Larsen RA et al (1996) Identification of patients with acute AIDS-associated cryptococcal meningitis who can be effectively treated with fluconazole: the role of antifungal susceptibility testing. Clin Infect Dis 22:322–328
133. Aller AL, Martin-Mazuelos E, Lozano F et al (2000) Correlation of fluconazole MICs with clinical outcome in cryptococcal infection. Antimicrob Agents Chemother 44:1544–1548
134. Dannaoui E, Abdul M, Michel-Nguyen A et al (2006) Results obtained with various antifungal susceptibility testing methods do not predict early clinical outcome in patients with cryptococcosis. Antimicrob Agents Chemother 50:2464–2470
135. Wheat LJ, Connolly P, Smedema M et al (2001) Emergence of resistance to fluconazole as a cause of failure during treatment of histoplasmosis in patients with acquired immunodeficiency syndrome. Clin Infect Dis 33:1910–1913
136. Walsh T, Annaissie EJ, Denning DW et al (2008) Treatment of Aspergillosis: clinical practice guidelines of the Infectious Diseases Society of America. Clin Infect Dis 46:327–360
137. Chapman SW, Dismukes WE, Proia LA et al (2008) Clinical practice guidelines for the management of Blastomycosis: 2008 update by the Infectious Diseases Society of America. Clin Infect Dis 46:1801–1812
138. Galgiani JN, Ampel NM, Blair JE et al (2005) Coccidioidomycosis. Clin Infect Dis 41:1217–1223
139. Saag MS, Graybill RJ, Larsen RA et al (2000) Practice guidelines for the management of cryptococcal disease. Clin Infect Dis 30:710–718
140. Wheat LJ, Friefeld AG, Kleiman MB et al (2007) Clinical practice guidelines for the management of patients with histoplasmosis: 2007 update by the Infectious Diseases Society of America. Clin Infect Dis 45:807–825
141. Kauffman C, Bustamante B, Chapman SW, Pappas PG (2007) Clinical practice guidelines for the management of patients with sporotrichosis: 2007 update by the Infectious Diseases Society of America. Clin Infect Dis 45:1255–1265

About the Editor

Gerri S. Hall, Ph.D. Dr. Hall's professional career has been focused on clinical microbiology: direct clinical activities of various areas such as bacteriology, mycobacteria, STD testing, and antimicrobial susceptibility testing of bacteria and fungi. Her laboratory performs clinically relevant patient care testing and research in these areas. She also teaches within the Department of Clinical Pathology as well as externally and is actively involved in teaching in the Cleveland Clinic Foundation Lerner School of Medicine of Case Western Reserve University.

G.S. Hall (ed.), *Interactions of Yeasts, Moulds, and Antifungal Agents:*
How to Detect Resistance, DOI 10.1007/978-1-59745-134-5,
© Springer Science+Business Media, LLC 2012

Index